GLENN TOOLE is Senior Tutor and Head of Biology at Woodhouse Sixth Form College, North Finchley. He has several years experience as an examiner of GCE O-Level Biology and he is currently an examiner for GCE A-Level Biology. He obtained an honours BSc from the University of London and he is a Member of the Institute of Biology.

Glenn Toole is co-author of the *A-Level Biology Course Companion* also published by Charles Letts Books Ltd.

Worked Examples

APPLIED MATHEMATICS, E. M. Peet, B.Sc.

BIOLOGY, A. G. Toole, B.Sc.

CHEMISTRY, J. E. Chandler, B.Sc. and
R. J. Wilkinson, Ph.D.

ECONOMICS, R. Maile, B.A.

GEOGRAPHY, C. Lines, M.Sc.

PHYSICS, G. M. George, B.Sc.

PURE MATHEMATICS, R. A. Parsons, B.Sc. and
A. G. Dawson, B.Sc.

PURE AND APPLIED MATHEMATICS,
R. A. Parsons, B.Sc. and A. G. Dawson, B.Sc.

GCE A-Level

Biology

A. G. Toole, B.Sc.

Published by Charles Letts Books Ltd
London, Edinburgh and New York

Published 1983 by Charles Letts Books Ltd
Diary House, Borough Road, London SE1 1DW

1st edition   1st impression
© Charles Letts Books Ltd
Made and printed by Charles Letts (Scotland) Ltd
ISBN 085097 552 2

Contents

Introduction

Examination techniques at A level

The successful A-level candidates are those who rapidly adapt from O level to the greater demands required of them at A level. In particular, success depends upon a mastery of examination techniques and broad experience of answering the various types of examination questions set at A level. The major difference between O- and A-level examination questions is the greater emphasis on understanding and application of knowledge at A level rather than learning and recall of factual information, which is common at O level. An A-level candidate is expected to appreciate the underlying principles, concepts and ideas behind a question, as much as the factual content associated with it. Compared with O level a much greater emphasis is placed on the ability to apply biological knowledge and principles to novel situations. The ability to interpret data in a variety of forms is required, and this is often deliberately obscure data which a candidate will never previously have encountered. In this way examiners are able to test understanding as opposed to rote learning. More open-ended questions occur at A level and candidates may not be required to give a specific answer, but rather to argue a case for or against a particular view. The ability to analyse data, experiments and information critically is necessary at A level as well as evaluating the accuracy of results and theories. Such skills require the candidate to reason objectively, maintain a balanced view and argue logically.

Approach to the various types of worked examples

There are no simple methods of mastering examination techniques as this is largely a matter of experience gained from practising questions and comparing answers with those expected by the person setting the question. There are however a number of basic rules:

1 Read all instructions carefully. Note especially the number of questions to be answered, any questions which are compulsory and instructions for the use of diagrams.

2 Read all questions carefully, preferably twice, and select possible questions to be attempted where a choice is available.

3 Always attempt the maximum number of questions required by the examiners.

4 Plan your time for answering according to the marks allocated. Do not spend more than the allocated time on a question.

5 For each question attempted, underline the most significant

words. It is important to refer back to the question during the writing of your answer.

6 Never include irrelevant material. This will not gain marks and will waste valuable time.

7 Diagrams should be included where they make a point clearly and save time. Do not repeat in prose, information shown on a diagram. All diagrams should be large and clearly labelled or annotated. Do not waste time shading large areas.

8 Make sure your writing is legible, answers are well ordered and spellings, especially of words specific to biology, are correct.

9 When all the required answers have been attempted read through them and check that all instructions have been complied with.

10 On multiple choice papers make sure you have answered every question. Normally no marks are deducted for incorrect guesses. If you have marked more than one response for a question neither will be marked as correct.

11 On highly structured questions the space allocated in the answer book may guide you in the length of your answer. Answer concisely and clearly.

Whether or not an examining board sets a separate practical examination, questions on theory papers are often based on practical work and experimental procedures. It is important that candidates do not separate these two aspects of their biological work, but take every relevant opportunity to answer questions using the information obtained from practical work. Theory papers are generally designed to test, by a variety of means, the following abilities:

1 recall of learned facts and principles.

2 application of learned facts and principles.

3 understanding and interpretation of data presented in a variety of forms, e.g. graphs or tables.

4 construction of hypotheses and design of experiments.

5 expression of information by means of prose, graphs, tables, drawings and diagrams.

6 evaluation of scientific information and presentation of logical, written arguments.

The questions in this book are designed to test each of the six abilities listed above, although where possible the abilities 2-6 have been emphasized because these are the ones candidates find most difficult and are the ones where specific guidance can be given. Certain topics, e.g. classification, however, do not lend themselves to being tested by other than ability 1 and examples in these chapters therefore primarily test recall.

All questions have been chosen after a thorough analysis of the past ten years' examination papers of the 10 examining boards in the United Kingdom. They represent a cross section of all the types of objective question set. As the book is intended to give worked examples, only questions which have a large element of objectivity, i.e. questions with a more or less definite answer, have been chosen. Essay-type questions with their scope for candidates to explore their own ideas and reveal their varied biological knowledge, have been avoided. It would be impossible, even if the author wished, to give precise information on the exact answer to such questions. Instead the emphasis has been to guide the reader on how to answer objective questions, what information to include and how to present it. Not only are the answers given, but the means by which the answers are obtained are also explored. Various pitfalls for the unwary and likely errors are pointed out. Where there are alternatives these are listed, and when feasible an order of priority is designated. Time allowances are provided to enable candidates to attempt the questions themselves and to provide some indication of the amount of detail required in the answer. The time allowances may appear too short in view of the extent of the answer that follows. This is because the answers include explanations and alternatives which would not be included when giving answers under examination conditions. .

Candidates will derive most benefit from the book if used throughout their A-level course. It will be most valuable if used to test their knowledge and understanding of each topic as it is studied. In addition it can be used as a revision aid in the final run up to the A-level examination. Candidates are advised to attempt the examples themselves before referring to the answers and explanations. Only in this way will they be able to isolate their own weaknesses and so be able to concentrate on improving them. It is hoped that in this way candidates will not only improve their biological knowledge but also their examination technique and by so doing greatly enhance their chances of success at A level.

Chapter 1   Classification

Example 1 (Time allowance 1 minute)
Below are 5 statements concerning the classification of organisms. Choose the one which is correct.

A   The scientific name of an organism refers to its class and species respectively.
B   Where an organism has a larval stage, this stage is given a different scientific name from the adult stage.
C   In a given phylum each class comprises the same number of orders.
D   In certain circumstances the scientific name of an organism can be changed.
E   All members of a species are morphologically identical.

Answer   D
The scientific name may be changed if, for instance, as a consequence of more thorough investigation an organism once thought to have belonged to one genus is found to be more closely related to another. Occasionally what was thought to be a distinct species is found to be the larval/juvenile stage of another species and so its scientific name may again be changed. A is wrong because the scientific name refers to the **genus** (not class) and the species respectively. All immature stages take the same scientific name as the adult, thus B is incorrect. Because the division of classes into orders is based on evolutionary relationships rather than taxonomic convenience the number of orders in the classes of a given phylum varies widely. This means C is inaccurate. E is incorrect because different environmental influences produce different phenotypes. Genotypic variation, not least that between males and females, also produces morphological variation within a species.

Example 2 (Time allowance 1 minute)
The wing of a housefly and of a bat are
A   both homologous and analogous.
B   analogous not homologous.
C   homologous not analogous.
D   neither homologous nor analogous.

Answer   B
An organ of one organism is said to be homologous with an organ

of another if both have a fundamental similarity of structure and/or position which is manifested especially during embryonic development, regardless of their functions in the adult. The housefly and bat wings have no such basic similarity and are therefore not homologous. Options A and C can therefore be discounted. An organ of one species is said to be analogous to an organ of another when both organs have the same function and when they are not homologous. The wings of a housefly and a bat are both used for flying but are not homologous and are therefore analogous. Option D may therefore be discounted and Option B is clearly correct.

Example 3 (Time allowance 1 minute)
For the following question determine which of the responses that follow are correct answers to the question. Give the answer A, B, C, D or E according to the key below:
A 1, 2, and 3 are correct
B 1 and 3 only are correct
C 2 and 4 only are correct
D 4 alone is correct
E 1 and 4 only are correct
In constructing a natural classification the following should be used:
1 comparative embryology
2 individual pedigrees
3 habitat differences between organisms
4 structural similarities between organisms

Answer E
Both comparative embryology and structural similarities between organisms are used in the construction of a natural classification. The pedigree of individuals is not used because it varies widely between individuals which nonetheless are capable of interbreeding and so belong to the same species. Habitat differences are frequently a consequence of environmental conditions and may even change during the lifetime of a single individual, especially where there is a larval stage. Natural classification is based upon more permanent genetic differences. Options 2 and 3 are therefore incorrect and 1 and 4 are correct.

Example 4 (Time allowance 5 minutes)
(a) For what purposes are organisms classified?
(b) Place the following list of taxonomic groups in ascending order beginning with the smallest unit of

---

> classification and finishing with the largest:
> class; phylum; species; order; family; genus.
> (c) What is meant by the term genus?

Answer

(a) (1) It is more convenient, especially when dealing with large volumes of information on organisms, to arrange it in a set order so that future reference to it is made easier.
(2) When an organism is given a single universally accepted name, scientists may communicate about it without risk of ambiguity.
(3) Classification indicates similarities and evolutionary relationships between members of one group.

(b) Species; genus; family; order; class; phylum

(c) The term 'genus' is derived from the Greek word 'genos' meaning race or stock and is a group of structurally or phylogenetically related species. It is the first of the two names given to an organism under the binomial system, i.e. the one shared with other closely-related organisms.

Example 5 (Time allowance 5 minutes)
(a) Explain why a virus can be considered as both living and non-living.
(b) List 4 differences between a typical plant cell such as a palisade mesophyll cell and a typical animal cell such as a smooth muscle cell.

Answer

(a) A virus is capable of taking over the biochemical pathways of a host cell and using them to produce new viruses, i.e. it is capable of reproduction, a fundamental process in living organisms. Outside the host cell it cannot reproduce itself and some viruses at least can be crystallized. These are features associated with non-living structures. A virus may therefore be considered as both living and non-living.

(b) Choose from:

Plant cell	Animal cell
Cellulose cell wall present in addition to the cell membrane	No cellulose cell wall present, only a cell membrane
Chloroplasts containing chlorophyll present	No chloroplasts or chlorophyll present
Lysosomes absent	Lysosomes present
Centrioles absent	Centrioles present
Starch grains present but no glycogen granules	Glycogen granules present but no starch grains
Central vacuole with cell sap	No central vacuole or cell sap

Chapter 2 The animal kingdom

Example 1 (Time allowance 5 minutes)
(a) State the phylum to which each of the following animals belong: a ciliate; an anuran; a bivalve; a polychaete.
(b) A hydroid such as *Hydra* and a jellyfish such as *Aurelia* are classified as coelenterates. State 3 reasons why they are classified in this phylum.
(c) List 3 ways by which a planarian and a hydroid could be distinguished.

Answer
(a) Protozoa; Chordata (Amphibia is a class, not a phylum); Mollusca; Annelida.
(b) Choose any 3 from:
    (i) Diploblastic (body divided into 2 layers, endoderm and ectoderm)
    (ii) Nematocysts present on tentacles
    (iii) Single internal cavity (gastro-vascular cavity or enteron)
    (iv) Single opening of the gut to the exterior
    (v) Radially symmetrical
(c) Choose any 3 from:

Planarian	Hydroid
Triploblastic	Diploblastic
Bilaterally symmetrical	Radially symmetrical
CNS present	CNS absent
Nephridial (excretory) system present	No specialized excretory system
Nematocysts absent	Nematocysts present
Cilia but no tentacles	Tentacles but no cilia
Protrusible pharynx	No pharynx
Photoreceptors present	Photoreceptors absent

Example 2 (Time allowance 10 minutes)
(a) Bilateral symmetry, triploblastic organization, segmentation, presence of a coelom and nephridia are characteristics of some animals.
    (i) State the phylum to which these animals belong.
    (ii) Give the name of the class of animals in this phylum that characteristically possess a clitellum.
    (iii) State 1 other way in which this class differs from others in this phylum.

(b) (i) Give the name of another phylum that possesses 4 of the characteristics listed in (a).
(ii) Which characteristic listed is **not** shown by this second phylum?
(iii) List 2 other ways of distinguishing between members of the two phyla in addition to any in the list of five in part (a).

Answer

(a) (i) Annelida
(ii) Oligochaeta
(iii) Oligochaeta means 'few bristles' and this class have no parapodia and only a few small chaetae in each segment (except first and last). The Polychaeta have many chaetae borne on parapodia and the Hirudinea have neither chaetae nor parapodia.
(b) (i) Arthropoda
(ii) Presence of nephridia
(iii) Choose from:

Annelida	Arthropoda
Closed blood system	Open blood system (haemocoel)
No jointed appendages	Jointed appendages present
Body muscle in continuous layers	Body muscles in separate bundles
Collagenous cuticle	Chitinous exoskeleton
Never compound, only simple eyes where present	Compound and/or simple eyes

Example 3 (Time allowance 1 minute)
To which of the following list of five groups of animals does a bilaterally symmetrical, coelomic, segmented organism with haemoglobin and a ventral nerve cord belong?
A insects
B platyhelminths
C annelids
D molluscs
E vertebrates

Answer C

A can be discounted because insects do not possess haemoglobin, B because platyhelminthes are not segmented or coelomic, nor do they possess haemoglobin. D is not possible because molluscs are unsegmented and haemocyanin, not haemoglobin, is the respiratory pigment when this is present. E is incorrect because vertebrates have a dorsal, not a ventral, nerve cord. Although not all annelids have haemoglobin, some do. The remaining features are shared by all annelids, therefore C is the correct answer.

Example 4 (Time allowance 5 minutes)
When Linnaeus first classified living organisms he grouped all worm-like animals together under the title 'Vermes'. Modern classifications divide the worms into a number of groups.
(a) Give the names of the 3 major phyla of worms.
(b) For each phylum list a single feature found in members of that phylum but not in the other two.
(c) About 65 species of segmented animals with jointed limbs and a tracheal system, but no exoskeleton are known to exist. Most of their relatives are known only as fossils.
   (i) Between which 2 phyla should these animals be placed?
   (ii) State the term often used to describe such animals.

Answer
(a) Platyhelminthes, Nematoda, Annelida.
(b) It is advisable to keep to features found in most, if not all, members of the phylum, rather than listing features found only in some members, therefore choose from:

Platyhelminthes	Nematoda	Annelida
Single gut opening	Pseudocoelomate	Coelom present
Gut caecae	Longitudinal muscle only	Segmented
		Blood system present

(c) (i) Annelida and Arthropoda **(ii)** Missing links

Example 5 (Time allowance 3 minutes)
What is meant by (a) triploblastic organization (b) coelomate animals (c) metameric segmentation?

Answer
(a) This means that the body is composed of three cellular layers, endoderm, mesoderm and ectoderm. These layers often become indistinct during the development of an organism and so only clearly show themselves in early embryonic stages.
(b) These are triploblastic animals in which the mesoderm is split into two layers separated by a fluid-filled space (the coelom).

14

(c)   This occurs when during embryonic development the body of certain animals becomes divided up into a series of similar segments, the number of which remains constant. Many of the contents of the body are repeated in each segment.

Example 6 (Time allowance 10 minutes)
Below is a jumbled list of 15 diagnostic characteristics of 3 different animals. For each animal there are 3 characteristics of the phylum and 2 characteristics of the class to which it belongs. Identify the phylum and class of each animal and state the letters of the characteristics that are typical of the phylum and class chosen.
(a)   Jointed appendages present
(b)   Unsegmented, but with a head bearing tentacles
(c)   Mantle present which secretes calcareous shell
(d)   Chitinous exoskeleton
(e)   Body divided into head, thorax and abdomen
(f)   Asymmetrical due to torsion
(g)   Spiny body
(h)   Tracheal system opening to the outside by way of spiracles
(i)   Body divided into head with tentacles, dorsal visceral mass and a ventral muscular foot
(j)   Compound eyes present
(k)   Radially symmetrical
(l)   Ambulacral grooves present
(m)   Water vascular system and tube feet present
(n)   Flattened star-shaped body
(o)   Radula present

Answer

Animal			Features
A	Phylum	Mollusca	b, c, i
	Class	Gastropoda	f, o
B	Phylum	Echinodermata	g, k, m
	Class	Asteroidea	l, n
C	Phylum	Arthropoda	a, d, j
	Class	Insecta	e, h

Example 7 (Time allowance 10 minutes)
(a)   Which class of chordates has all its members with skin permeable to water?
(b)   Give 5 reasons why birds are a successful group.
(c)   Why are fish and mammals in the same group?
(d)   List 3 differences between a lizard and a frog.

Answer
(a) Amphibia
(b) Choose from: (the earlier features are more important)
    1   Feathers for flight/insulation allowing birds to maintain a high body temperature
    2   Endothermic (homoiothermic)/high metabolic rate
    3   High degree of parental care/nesting
    4   Chalky waterproof egg prevents desiccation on land so that birds can reproduce without returning to water
    5   Forelimbs developed into wings to allow flight
    6   Excretion of uric acid conserves water
    7   Internal fertilization
    8   Hollow bones reduce weight and make flight easier
    9   Air sacs used in breathing
   10   Good eyesight (rapid accommodation)
(c) Both possess the following chordate vertebrate features:
    1   A dorsal hollow nerve cord
    2   A cranium
    3   A notochord (skeletal rod) in the embryo—vertebral column in the adult
    4   Gill clefts/pharyngeal gill slits/visceral clefts at some stage in the life history
    5   Metamerically segmented post-anal tail is present at some stage in the life history
    6   A ventral heart which pumps blood backwards in the dorsal vessel and forwards in the ventral vessel
    7   Jaws are present
    8   Kidneys are present
(d) Choose from:

Lizard	Frog
No webbed feet	Webbed feet
Scales present	Scales absent
Five digits on forelimb	Four digits on forelimb
Hindlimbs and forelimbs roughly comparable in size and strength	Hindlimbs considerably longer and more powerful than fore-limbs
Twelve pairs of cranial nerves	Ten pairs of cranial nerves
Ventricle of heart partly divided	Ventricle of heart not divided at all
Tongue attached posteriorly	Tongue attached anteriorly
Eyes lower on head	Eyes placed high up on head
Jacobson's organ present	Jacobson's organ absent

Chapter 3   The plant kingdom

Example 1 (Time allowance 4 minutes)
(a)   Give 2 reasons why bacteria are successful.
(b)   State 2 differences between prokaryotic and eukaryotic cells.

Answer
(a)   Choose from the following:
1   Bacteria reproduce very rapidly and thus colonize new areas quickly. Rapid reproduction means mutant types are more often produced and this produces the evolutionary potential to adapt to new habitats and changing conditions.
2   Their small size means they have small nutritional requirements and are easily dispersed.
3   They may use a diverse range of nutrients and so survive in almost all environments.
4   Sexual exchange of genetic material between different types of bacteria gives genetic variety and a consequent ability to adapt to new environments.
(b)   Choose from:

Prokaryotic	Eukaryotic
No nuclear membrane	Nuclear membrane present
No membrane-bounded intracellular organelles	Membrane-bounded organelles, e.g. chloroplasts, mitochondria are present
Genetic material is a strand of DNA	Chromosomes (DNA + protein) make up genetic material

Example 2 (Time allowance 3 minutes)
(a)   (i)   Name **one** feature of *Mucor* that suggests that it is a member of the plant kingdom.
    (ii)   Name **one** feature of *Mucor* that contrasts with the general characteristics of plants.
(b)   In view of your answer to (a) why are fungi usually considered to be members of the plant kingdom?

Answer
(a)   **(i)**   Choose from:
    1   There is a large central sap-filled vacuole.
    2   The cell membrane is surrounded by a cell wall.
    3   Growth is largely confined to growing points at the hyphal tips.

(ii) Choose from:
1 The absence of chlorophyll and nutrition is therefore heterotrophic.
2 The absence of cellulose (the cell wall is made of chitin).
3 Glycogen and/or oil is stored in contrast to starch.
(b) It was once thought that fungi evolved from the algae. Although this view is not so generally accepted today it is the historical reason for classifying fungi as plants.

Example 3 (Time allowance 3 minutes)
The following drawing represents a whole plant.

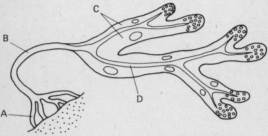

(a) State the names of the structures labelled A–D.
(b) This plant belongs to which major group?
(c) In what habitat would you expect to find this plant?

Answer
(a) A holdfast B stipe C thallus D mid-rib
(b) Algae (Thallophyta would probably be allowed).
(c) The littoral zone of a rocky sea shore.

Example 4 (Time allowance 4 minutes)
Bryophytes and Pteridophytes both show alternation of generations.
(a) Explain what is meant by this term.
(b) How are the life cycles of these groups affected by weather conditions?

Answer
(a) The life cycle comprises two generations which alternate during a plant's lifetime. One, the gametophyte, reproduces sexually by means of gametes and the other, the sporophyte, reproduces asexually by means of spores.
(b) The spores (asexual stage) are only released under dry conditions whereas the sperms (sexual stage) are transferred in a film of water (i.e. wet conditions).

18

Example 5 (Time allowance 8 minutes)

(a) State 4 similarities between gymnosperms and angiosperms that cause them to be classified together as Spermatophyta.

(b) State 4 differences between gymnosperms and angiosperms that result in them being classified in separate sub groups of the Spermatophyta.

(c) State 3 features by which monocotyledons may be distinguished from dicotyledons.

Answer

(a) Choose from:

1 a dominant sporophyte with a much reduced gametophyte
2 the embryo sac is enclosed in an ovule
3 separate male (pollen grains) and female (embryo sac) spores
4 presence of a pollen tube
5 development of fertilized embryo sac into a seed
6 vascular tissues comprising phloem and xylem

(b) Choose from:

	Gymnosperms	Angiosperms
1	No fruits formed	Fruits formed
2	Ovules unprotected	Ovules protected
3	No style or stigma	Style and stigma present
4	No flowers (usually), cones	Flowers, no cones
5	No xylem vessels, only tracheids	Tracheids and vessels in the xylem
6	No companion cells in the phloem	Companion cells present in the phloem

(c) Choose from:

	Monocotyledons	Dicotyledons
1	One cotyledon or seed leaf	Two cotyledons
2	Generally parallel-veined leaves	Generally net-veined leaves
3	Floral parts in threes or multiples of three	Floral parts in groups of 4 or 5 or multiples thereof
4	Vascular cambium mostly lacking	Vascular cambium in all forms with secondary growth
5	Vascular bundles scattered throughout the stem	Stem vascular bundles arranged in the form of a cylinder

Example 6 (Time allowance 10 minutes)
For each of the following descriptions state the precise name of the group to which each plant belongs and give a named example from that group.
(a) A filamentous, non-green saprophyte with no septa (cross-walls).
(b) A seed-producing plant showing alternation of generations with the gametophyte generally reduced. The ovule is unprotected and the xylem has no vessels.
(c) A motile green unicell.
(d) A plant showing alternation of generations with the gametophyte differentiated into a simple stem, leaves and rhizoids. It bears the sporophyte which derives nourishment from it.
(e) A multicellular green plant whose spore-producing organs are borne on the main photosynthetic structures.
(f) A plant showing alternation of generations in which the gametophyte generation is reduced and independent of the sporophyte. The sporophyte is differentiated into roots, stem and leaves.

Answer
(a) Group Phycomycetes (other fungal groups have cross-walls)
Example *Mucor, Phytophthora, Pythium, Saprolegnia*, etc.
(b) Group Gymnospermae (other spermatophyta have xylem vessels)
Example *Pinus, Taxus, Larix*, etc.
(c) Group Chlorophyta
Example *Chlamydomonas, Euglena*, etc.
(d) Group Bryophyta (as some liverworts and all mosses are leafy it is not possible to be more precise)
Example *Funaria, Polytrichum, Mnium*, etc.
(e) Group Filicales (in other pteridophyte groups spores are borne on non-photosynthetic shoots (Equisetales) or in the axils of leaves (Lycopodiales)
Example *Dryopteris, Pteridium*, etc.
(f) Group Pteridophyta (cannot be more specific, unlike (e))
Example *Lycopodium, Selaginella, Equisetum* + any examples from (e)

Chapter 4   Basic biochemistry

Carbohydrates

Example 1 (Time allowance 20 minutes)
(a) IIow would you detect the presence of (i) soluble starch and glucose in a solution of both substances? (ii) sucrose in a solution of sucrose and glucose?
(b) Give the name of a sugar other than sucrose or glucose and state precisely where it occurs naturally and the role it performs there.

Answer

(a) (i) Two separate samples of the solution should be taken, each of about $2 \, cm^3$. To the first should be added a few drops of iodine. If the iodine changes colour from yellow-orange to blue-black then the presence of starch is confirmed. To the second sample should be added an equal volume of Benedict's or Fehling's reagent. If the Benedict's (Fehling's) reagent changes colour from pale blue to orange/brown/ brick red on boiling, then a reducing sugar is present, in this case glucose.

(ii) Glucose is a reducing sugar whereas sucrose is a non-reducing sugar. Although there are different tests for each of these sugars it is not a simple matter of testing for each in turn in a similar way to that shown above. The non-reducing sugar test will give a positive result in the presence of a reducing sugar. The easiest method of detecting these two sugars is to take two equal samples of the solution, carry out the reducing sugar test on one and the non-reducing sugar test on the other and then compare the amounts (or possibly the colour) of the precipitate formed in each case.

The reducing sugar test ($2 \, cm^3$ of the test solution + $2 \, cm^3$ of Benedict's (Fehling's) reagent boiled together) should give a small amount of orange/brown precipitate because only the glucose, not the sucrose will reduce the Benedict's (Fehling's) reagent. In the non-reducing sugar test, $2 \, cm^3$ of the test reagent is first boiled with $1 \, cm^3$ of dilute hydrochloric acid for 5 minutes to hydrolyse the sucrose into glucose and fructose. The solution should then be neutral- ized with 10% sodium hydroxide (test with indicator paper). To the resulting solution should be added $2 \, cm^3$ of Benedict's (Fehling's) reagent and the mixture returned to the boil. The mixture now contains glucose and fructose from the hydrolysis of the sucrose (in addition to the original glucose) all of which will reduce the Benedict's (Fehling's) reagent.

The precipitate in this second test should therefore be darker and greater in quantity indicating that sucrose as well as glucose was present in the original mixture. The success of this method depends on the two original samples being of exactly equal volume and their being heated with the same amount of Benedict's (Fehling's) at the same temperature for the same length of time.

(b) Choose one from:

Sugar	Location	Function
Lactose	Mammalian milk	Nutrition for sucklings
Fructose	(i) Nectar in flowers	Attraction of insects as a reward for pollinating
	(ii) Many succulent fruits	Attraction of animals to aid seed dispersal
	(iii) Bee's honey	Food store
Ribose	RNA of all cells, especially in the nucleolus	Essential for protein synthesis

---

Example 2 (Time allowance 1 minute)
The general formula for a disaccharide is

  A $C_{12}H_{22}O_{11}$ C $C_3H_6O_3$ E $C_6H_{12}O_6$
  B $C_5H_{10}O_5$ D $C_{12}H_{24}O_{12}$

Answer A

Disaccharides are formed from two monosaccharides by the loss of a water molecule. Monosaccharides have the general formula $C_6H_{12}O_6$ (thus eliminating option E) and therefore two such molecules have the formula $C_{12}H_{24}O_{12}$ (thus eliminating option D). If the water molecule (H_2O) is removed, the disaccharide formula is seen to be $C_{12}H_{22}O_{11}$ (Option A). Options B and C are pentose and triose monosaccharides respectively.

Fats

Example 3 (Time allowance 10 minutes)
(a)   What is the general formula for a fatty acid?
(b)   Show how a fatty acid may be converted into a triglyceride.

22

(c) What is the general name for the reaction you have described in (b)?

(d) Many plants contain oils which are valuable sources of food.
(i) From which part of the plant are they often extracted?
(ii) Explain why you think the plant uses oil as a food source in these parts.
(iii) Which solvent would you use in a school laboratory to extract the oil?

(e) (i) Where do you find waxes in plants?
(ii) What are their functions?

Answer

(a) $R(CH_2)_nCOOH$, where R is an alkyl group such as CH_3 or C_2H_5.

(b) Three fatty acid molecules combine with a single glycerol (propantriol) in a condensation reaction to form a single triglyceride (propantriol tri-ester) molecule.

$$R(CH_2)_n\ COOH + HO - CH_2 \rightarrow R(CH_2)_nCOOCH_2$$
$$R(CH_2)_n\ COOH + HO - CH \rightarrow R(CH_2)_nCOOCH + 3H_2O$$
$$R(CH_2)_n\ COOH + HO - CH_2 \rightarrow R(CH_2)_nCOOCH_2$$

fatty acid + glycerol $\longrightarrow$ triglyceride
(alkanoic (propantriol) (propantriol tri-ester)
acid)

(c) Condensation reaction.

(d) (i) The seeds.
(ii) Compared with protein ($22 \cdot 2kJ/g$) or carbohydrate ($17 \cdot 2kJ/g$), oils ($38 \cdot 5kJ/g$) contain about twice as much energy for the same weight of material. The seed requires an energy source to permit growth during germination in order that the plant may establish leaves adequate for it to begin photosynthesising and so produce its own food. However, because all seed-producing plants are incapable of locomotion, the dispersal of seeds necessary to prevent overcrowding and intraspecific competition for light, water, etc. must be carried out by some external agent such as the wind or insects. In either case the seed must have a very small mass to permit wide dispersal, and yet carry adequate food. Oils are by far the best of the available materials satisfying these criteria.
(iii) An alcohol, e.g. ethanol or propan-2-ol.

(e) (i) In the cuticle which covers many parts of the plant, especially the leaves.
(ii) To form a waterproof barrier and so prevent excessive evaporative loss of water.

23

Proteins

Example 4 (Time allowance 5 minutes)
(a) Give the basic structure of an amino acid.
(b) Name the two groups which occur in all amino acids.
(c) Show in a diagram how two amino acids combine to form a dipeptide.
(d) What are the differences between globular and fibrous proteins?

Answer

(a)

(Where R may be a variety of radicals ranging from H to complex ring structures)

(b) Carboxyl ($-COOH$) and amino ($-NH_2$).

(c)

(d) Globular proteins are chains of amino acids formed into complex three dimensional structures due to the formation of weak electrostatic (hydrogen) bonds formed between carboxyl and amino radicals. Fibrous proteins consist of long parallel chains with cross links. Globular proteins are soluble and have largely metabolic functions (e.g. all enzymes and some hormones are globular proteins). Fibrous proteins are insoluble and have largely structural functions (e.g. collagen in cartilage, keratin in hair, feathers and hooves).

Example 5 (Time allowance 2 minutes)
Using the following list, relate one of the types of reaction

with each of the examples listed below. Each letter may be used once, more than once or not at all.

A condensation   B hydrolysis   C reduction   D oxidation

  (i)   Amino acids $\longrightarrow$ peptides
 (ii)   Fats $\longrightarrow$ fatty acids and glycerol
(iii)   Glucose $\longrightarrow$ maltose
(iv)   Sucrose $\longrightarrow$ glucose and fructose
 (v)   Cytochrome $Fe^{3+} \longrightarrow$ Cytochrome Fe^{2+}

Answer

Fats, proteins, disaccharides and polysaccharides are built up from small units by condensation reactions. Conversely these products are broken down into their component units by hydrolysis reactions. From this information it can be seen that (i) and (iii) are condensation reactions (Answer A) and (ii) and (iv) are hydrolysis reactions (Answer B). In (v) Fe^{3+} converted to Fe^{2+}, i.e. it has become less electropositive by gaining an electron. This is hence a reduction reaction (Answer C).

Summary: (i) A   (ii) B   (iii) A   (iv) B   (v) C

Nucleic acids

Example 6 (Time allowance 10 minutes)

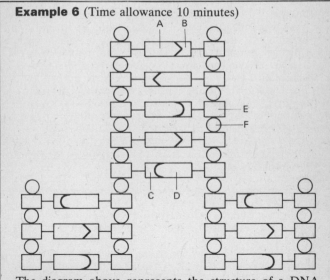

The diagram above represents the structure of a DNA molecule.

(a)   (i)   To which group of chemicals do A, B, C and D belong?
     (ii)   Within this group there are two sub groups. What are they and to which sub groups do each of A, B, C and D belong?
     (iii)   Deduce the names of A, B, C and D as accurately as possible.
     (iv)   Name E and F.

(b)   What is the importance of the process shown in the lower part of the diagram?

(c)   (i)   Name three types of RNA involved in protein synthesis.
     (ii)   Which chemical absent in DNA is found in RNA?

Answer

(a)   **(i)**   They are all nitrogen bases.

     (ii)   The sub groups are purines and pyrimidines. The purines comprise a double ring structure and pyrimidines a single ring. Purines are larger molecules than pyrimidines and it is reasonable to assume therefore that A and D are purines and B and C are pyrimidines.

     (iii)   It has been established in (a) (ii) that A and D are purines. These are therefore adenine and guanine although it is not possible to say which is which. Likewise B and C (the pyrimidines) are cytosine and thymine, but again which is which cannot be deduced from the information given.

     (iv)   E is the five-carbon sugar deoxyribose and F is phosphoric acid.

(b)   The DNA is separating into its two component strands. It may either be replicating itself in order to form two DNA strands each identical to the original or forming a RNA molecule that may be used as a template in the synthesis of proteins.

(c)   **(i)**   1 messenger RNA 2 transfer RNA 3 ribosomal RNA
     (ii)   The nitrogen base uracil (which replaces the thymine found in DNA).

Enzymes

Example 7 (Time allowance 8 minutes)
Sketch the graphs which would be obtained in an experiment to measure the rate of enzyme action at different
(i)   enzyme concentrations (using excess substrate).
(ii)   substrate concentrations (using a fixed quantity of enzyme).
Explain the shape of your graph in both cases.

Answer

(i) If the 'lock and key' theory of enzyme action is accepted then the rate of enzyme action is dependent upon the rate at which enzyme and substrate molecules come into contact with each other. Any doubling of the number of enzyme molecules present will double the chances of an individual enzyme molecule contacting a substrate molecule. The rate of reaction therefore doubles. The graph therefore rises linearly and provided the substrate is in excess there is no 'tailing off' of the graph.

(ii) In the same way as (i) the increase in substrate concentration increases the chances of enzyme and substrate molecules colliding so the graph rises linearly. However, since the enzyme concentration is fixed, there comes a point where the amount of enzyme limits the rate of reaction. At this point all enzyme molecules are simultaneously acting on substrate molecules and a further increase in substrate concentration has no effect on the rate of reaction. The graph therefore 'tails off' to a plateau. It never falls because, having once acted on a substrate molecule, each enzyme molecule is free to be used on further substrate molecules.

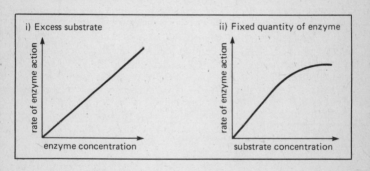

Example 8 (Time allowance 20 minutes)

Dichlorophenolindophenol (DCPIP) is an indicator which is blue when oxidized and turns colourless when reduced. The liquid extracted from some mammalian tissue was tested for the presence of the enzyme succinic dehydrogenase using DCPIP. The enzyme catalyses the transfer of hydrogen from succinic acid to DCPIP thereby reducing it.

Four experiments were set up as follows:

Contents	Tube 1	Tube 2	Tube 3	Tube 4
Succinic acid	+	−	+	+
DCPIP	+	+	+	+
Tissue extract	+	+	−	−
Boiled tissue extract	−	−	−	+
Distilled water	+	+	+	+

+ = present   − = absent

The contents of all tubes were covered with a thin layer of oil and placed in a water bath at 35°C. After a few minutes only Tube 1 was colourless, all the rest being blue.
(a)   For each tube in turn explain why it was necessary to use the combinations of contents used.
(b)   What conclusions can be drawn from the results?
(c)   Why was oil added to the contents of each tube?
(d)   Why were the tubes placed in a water bath at 35°C.

Answer

(a)   **Tube 1**   This includes the tissue extract which contains the succinic dehydrogenase, succinic acid from which the enzyme removes hydrogen and DCPIP to which the hydrogen is transferred. The DCPIP is therefore reduced and becomes colourless. The distilled water is to make up the volume to a constant amount in all tubes so these can be compared. This is necessary because the contents of each tube vary in quantity according to how many substances are present. Tube 1 is therefore the experimental tube containing all the necessary materials for the reaction.

Tube 2   This has similar contents to tube 1 except the succinic acid is replaced by distilled water. There is no change in the colour of DCPIP, thus showing that the reaction requires succinic acid (in this case to act as a donor of hydrogen). Tube 2 is therefore a control showing the necessity for succinic acid.

Tube 3   This is the same as tube 1 except that the tissue extract is replaced by distilled water. There is no change in the colour of the DCPIP, indicating that the extract contains some essential substance for the reaction to occur (in this case it contains the enzyme succinic dehydrogenase that catalyses the removal of hydrogen from succinic acid). Tube 3 is hence a control to show the necessity for tissue extract.

Tube 4   This is the same as tube 1 except that the tissue extract is boiled. Boiling denatures any enzymes in the extract thus rendering them non-functional. The fact that the DCPIP is not changed indicates that the essential factor in the extract is rendered useless by boiling. Tube 3 is hence a control providing additional evidence that the extract contains an essential enzyme (succinic dehydrogenase).

(b)   The fact that the reduction of DCPIP only occurred in tube 1 shows that the tissue extract contains an enzyme (probably succinic dehydrogenase) that catalyses the transfer of hydrogen from succinic acid to DCPIP.

(c)   As the DCPIP is reduced it becomes colourless. The presence of oxygen, however, reoxidises the DCPIP immediately and causes it to revert to its blue colour. The oil forms a film that considerably reduces the rate at which oxygen diffuses into the solution. Any oxygen already dissolved in the contents initially reoxidises the DCPIP, but this is quickly used up and because the oil inhibits further entry of oxygen the reaction can proceed normally.

(d)   The water bath helps to maintain a constant temperature because water has a large heat capacity and so buffers changes in external temperature. The temperature is kept at 35°C because the enzyme is in mammalian tissue extract. Most mammals have a constant body temperature in the region of 35-40°C. Most mammalian enzymes work efficiently within this temperature range and therefore 35°C is used in the experiment.

Analytical techniques

Example 9 (Time allowance 8 minutes)

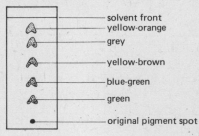

Above is a diagram of the results obtained when a spot of green plant leaf extract was loaded on a piece of filter paper which was then suspended with its lower end in a suitable solvent.

(a)   In what way could the extract have been obtained?

(b)   Why does the extract appear as a range of colours when separated on the filter paper?

(c)   How could the Rf value for any given pigment be calculated?

Answer

(a)   The leaves could have been removed from a suitable plant such as the stinging nettle (*Urtica dioica*). The leaves could be ground up using a mortar and pestle with a little sand and a suitable solvent such as acetone. The resultant liquor could be filtered through filter paper to give a dark green extract.

(b)   The extract is not a single substance but a mixture of many pigments each of which plays a specific role in photosynthesis. The following pigments commonly occur:

chlorophyll a (blue-green)
chlorophyll b (green)
carotene (yellow/orange)
xanthophyll (yellow/brown)
phaeophytin (grey)

Each pigment has a different solubility in the solvent and so each is moved a different distance up the filter paper. The most soluble is moved furthest, the least soluble is moved the shortest distance.

(c)   The Rf value is a means of helping to identify each separated pigment. For each pigment it is calculated as follows:

$$\frac{\text{distance moved by pigment spot}}{\text{distance moved by solvent front}}$$

(Both distances are measured from the original pigment spot)

Chapter 5   Cells and tissues

The ultrastructure of the cell

Example 1 (Time allowance 12 minutes)
Study the diagram below and then answer the questions that follow.

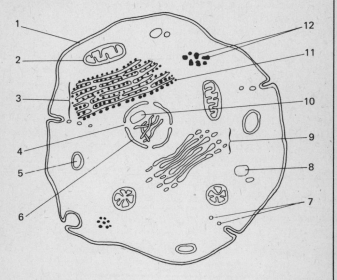

(a)   Label parts 1-12.
(b)   State the function of the part labelled 2.
(c)   Describe the events that link structures 10, 11 and 12.
(d)   What is the approximate magnification of this drawing?

Answer

(a)

1	plasma membrane	7	pinocytic vesicles
2	mitochondrion	8	oil droplet
3	rough endoplasmic reticulum	9	Golgi apparatus
		10	nucleolus
4	nuclear pore	11	ribosome
5	lysosome	12	glycogen granules
6	nuclear membrane		

(b)   This is a mitochondrion which is the centre of energy production within the cell. The invaginations of the inner

31

membrane (cristae) have attached to them the enzymes involved in the Krebs' (citric acid) cycle and the electron (hydrogen) carrier system.

(c) Structure 10 is the nucleolus which contains RNA. One type of RNA called messenger RNA, leaves the nucleus and enters the cytoplasm of the cell. It comprises a sequence of nitrogen bases that have been determined by the sequence of such bases on an appropriate section of DNA within the nucleus. The messenger RNA wraps itself around some ribosomes (structure 11). The sequence of nitrogen bases ultimately determines the sequence of amino acids in a protein which is synthesized. This protein could be an enzyme that is involved in the production of glycogen (structure 12), the main carbohydrate store of an animal cell.

(d) The best means of calculating this is to take an organelle whose actual size is known. The best organelle is usually a mitochondrion because its size, although variable, in length, is reasonably constant in diameter. The diameter is usually 0·75 μm. If the diameter of a few mitochondria are measured an average can be found. In this case it is near enough 5·0 mm (it is a good idea to approximate to a figure that will make the calculation reasonably easy).

$$\text{Actual size} = 0·75 \ \mu m$$
$$\text{Size drawn} = 5·0 \ \text{mm}$$
$$\text{Therefore magnification} = \frac{5·0 \ \text{mm}}{0·75 \ \mu m} = \frac{5000 \ \mu m}{0·75 \ \mu m}$$
$$(1 \ \text{mm} = 1000 \ \mu m)$$
$$= \textbf{6666} \text{ times approximately}$$

Example 2 (Time allowance 1 minute)
Choose the one correct response to the following statement. Cell membranes are:

A   about 8 nm thick and consist of two layers of phospholipid sandwiched between two layers of protein.

B   about 12 nm thick and consist of one layer of phospholipid sandwiched between two layers of protein.

C   about 8 nm thick and consist of two layers of protein sandwiched between two layers of phospholipid.

D   less than 5 nm thick and consist of one layer of phospholipid sandwiched between two layers of protein.

E   more than 12 nm thick and consist of two layers of phospholipid sandwiched between two layers of protein.

Answer A

The membrane thickness is too large in B and E and too small in D. In addition there is only one, rather than two, phospholipid layers in B and D. In C the number and the thickness of the layers is correct but the protein layers are sandwiched between phospholipid layers rather than the other way round. Only in A are all the facts correct.

Example 3 (Time allowance 5 minutes)
(a) Explain why lysosomes may be called 'suicide bags'.
(b) Why are lysosomes more frequent in phagocytic cells than in other cells?

Answer

(a) One function of lysosomes is to destroy old and worn out cells. The lysosomes contain lytic enzymes which may be released into the cell thus breaking it down and destroying it. This self-digestion (autolysis) by the cells' lysosomes is the reason for them sometimes being called 'suicide bags'.

(b) Phagocytic cells engulf foreign material either for food as in *Amoeba*, or in order to remove debris or bacteria, as in leucocytes. In either case the ingested material must be broken down and digested without destroying the phagocytic cell. The part to be destroyed is surrounded by a membrane and lysosomes discharge their lytic enzymes into this sac, digesting the contents, the products of which are absorbed into the cytoplasmic matrix. The number of foreign particles is greater in phagocytic cells and consequently lysosomes are more frequent in these.

Example 4 (Time allowance 1 minute)
Which of the following cells would most probably contain the greatest number of Golgi bodies?
A muscle cell
B secretory cell
C nerve cell
D white blood cell
E epithelial cell

Answer B

The function of Golgi bodies is thought to be concerned with the addition of carbohydrate to proteins produced by the endoplasmic reticulum. The glycoproteins so produced are contained in vesicles pinched off from the Golgi body. These vesicles move to the cell membrane and secrete their contents to the outside. Golgi bodies are therefore more numerous in secretory cells.

Animal tissues

Example 5 (Time allowance 4 minutes)
The following table relates to four different tissues.
Provide the missing information by giving appropriate
answers for each of the letters in the table.

Name of tissue	Main property of tissue	Location of tissue
A	Contracts semi-powerfully but is never fatigued	B
Smooth muscle	Controlled by autonomic nervous system	C
White fibrous connective tissue	D	E
F	G	At joints

Answer

A cardiac muscle
B heart
C wall of alimentary canal/blood vessels/uterus/urinary bladder
D strong and inelastic
E attached to bones
F yellow elastic connective tissue
G elastic and tough

Example 6 (Time allowance 7 minutes)
(a) What is a tissue?
(b) For each of the following tissues state two places in the
      mammalian body where it is commonly found and in
      each case state its importance: (i) ciliated
      epithelium (ii) squamous epithelium.

Answer
(a) A tissue is a collection of histologically similar cells grouped
together to perform one particular function.
(b) (i) This is found in the respiratory tract (trachea and
          bronchi) where it is responsible for the movement of
          dirt-laden mucus from the alveoli to the pharynx. It is

34

important in preventing the accumulation of this mucus in the alveoli and the consequent blockage of the respiratory tract.

It is also found in the oviducts (fallopian tubes) where it is important in wafting the non-motile ovum down towards the uterus where, if fertilized, it implants and develops into the foetus.

(ii) This lines the alveoli of the lungs where its thinness ($2 \cdot 0$ μm thick) is of particular importance in allowing diffusion of oxygen into the blood and carbon dioxide from the blood into the alveoli.

Squamous epithelium also lines blood capillaries in the tissues. Again its thinness is important in allowing diffusion of materials in and out of the blood, e.g. oxygen, glucose, amino acids and mineral salts may diffuse from the blood into the tissues and carbon dioxide and urea may diffuse in the reverse direction.

Plant tissues

> **Example 7** (Time allowance 15 minutes)
> (a) How does the structure of a parenchyma cell differ from a leaf epidermal cell?
> (b) For each of the following plant tissues give the name of a tissue in a mammal that performs the same function
>   (i) a leaf epidermis   (ii) xylem tracheids.
> (c) Comment on the differences between the comparable plant and animal cells given in (b).

Answer

	Parenchyma cell	Leaf epidermal cell
1	Approximately isodiametric in shape	Flat and thin in shape
2	Large vacuole	Small vacuole
3	Cytoplasm comprises a lesser percentage of the cell by volume	Cytoplasm comprises a greater percentage of the cell by volume
4	Thin cellulose wall	Thick outer wall of cellulose—other walls thin
5	Cutin absent on cell wall	Cutin present on outer cell wall
6	Storage granules, e.g. starch common	Storage granules rare
7	Chloroplasts sometimes occur	Chloroplasts rare, if at all

(b) (i) Stratified epithelium of the skin   (ii) Compact bone

(c) **(i)** Some differences between a leaf epidermis and stratified epithelium are no more than basic plant/animal cell differences, e.g. the presence in a leaf epidermal cell of cellulose, a single central vacuole and starch, all of which are absent in a stratified epithelium. Other differences however are a result of the fundamental functional differences between the two tissues. The leaf epidermis needs to be waterproof and therefore produces cutin and yet it must allow light for photosynthesis to pass to the cells beneath; it is therefore thin (one cell thick) and transparent. The stratified epithelium by contrast has the main role of protecting against mechanical abrasion although it too must be waterproof to some extent. It is therefore many cells thick, opaque and contains keratin which is resistant to abrasion and waterproof.

(ii) Apart from the basic plant/animal cell differences listed above, the differences between xylem tracheids and compact bone are few because both have one major function in common—that of support. Xylem tracheids, unlike bone, are however dead and therefore have no nucleus or cytoplasm. Xylem tracheids also conduct water and minerals and are consequently hollow with pits, whereas bone, although it has Haversian canals within it, is largely a solid tissue. The strengthening material in xylem tracheids is lignin which is also waterproof. In bone calcium salts are used to give rigidity.

Example 8 (Time allowance 6 minutes)
Below is a diagram of a transverse section through the stem of an herbaceous dicotyledon.

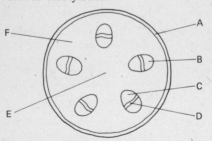

(a) Identify parts A-F.
(b) State the names of the cells in a stem which are
   (i) used as the main store of starch.
  (ii) have thickened lignified walls.
 (iii) transport sugars.

(c) Using the letters on the diagram state the regions where you would find:
  (i) tracheids. (ii) sieve tubes.
  (iii) cell division. (iv) dead cells.

Answer

(a) A   epidermis   D   cambium
     B   phloem   E   pith
     C   xylem   F   cortex
(b) (i) Parenchyma cells (ii) Xylem vessels and tracheid cells; sclerenchyma cells (iii) Phloem sieve tubes
(c) (i) C (ii) B (iii) D (iv) C

The passage of substances in and out of cells

> **Example 9** (Time allowance 1 minute)
> For the following question determine which of the responses that follow are correct. Give the answer A, B, C, D or E according to the key below:
> A 1, 2 and 3 are correct
> B 1 and 3 only are correct
> C 2 and 4 only are correct
> D 4 alone is correct
> E 1 and 4 only are correct
> Active transport:
> 1 proceeds at the same rate for all molecules.
> 2 is unaffected by changes in temperature.
> 3 continues to occur in the presence of cyanide.
> 4 may occur against a concentration gradient.

Answer D

Only statement 4 is accurate. Statement 1 is incorrect because the rate of uptake by active transport depends on whether the substance has its own carrier, or competes with other molecules for a carrier (as well as other factors). Statement 2 is incorrect because active transport requires energy which is supplied by respiration. Factors such as changes in oxygen concentration and temperature which affect the rate of respiration will therefore likewise affect the rate of active transport. Statement 3 is incorrect because cyanide is a respiratory inhibitor. In its presence respiration and hence energy production and active transport cease.

> **Example 10** (Time allowance 15 minutes)
> A portion of the ileum removed from a freshly killed mouse was turned inside out and filled with an oxygenated solution

of saline and sugars. It was tied at both ends and suspended in a solution of identical composition to that inside the sac. It was incubated at 37°C for two hours after which the concentration of sugars inside and outside the sac was measured. The results are shown below.

(a) By what process do you think L-glucose is being absorbed?
(b) Give two pieces of evidence that support your view.

Sugar in the solution	Conc (g/l) of sugar outside the sac	Conc (g/l) of sugar inside the sac
L-glucose	1·1	6·3
D-glucose	3·5	3·5
L-galactose	1·8	5·7
L-glucose (in deoxygen-ated solution)	3·5	3·5
L-galactose (in deoxygen-ated solution	3·7	3·7
L-glucose and L-galactose	1·4 (glucose) 3·0 (galactose)	6·0 4·5

(c) What happens to the rate of L-glucose and L-galactose absorption in the presence of each other as compared to their rate of absorption when separated?
(d) How might you account for the differences mentioned in (c)?
(e) Why is D-glucose not actively absorbed?

Answer

(a) Active transport.

(b) (i) After two hours L-glucose is concentrated inside the sac almost six times more than it is outside. It must therefore have been absorbed against the concentration gradient. This is a feature of active transport.

(ii) In the absence of oxygen the concentrations of L-glucose inside and outside the sac are equal. This means no concentrating of L-glucose has occurred in the absence of oxygen. Active transport is an energy-consuming process which only occurs in the presence of oxygen.

(c) The rate of L-glucose absorption is reduced very slightly while that of L-galactose is reduced more markedly (by about one half).

(d) If L-glucose and L-galactose use the same carrier for their absorption by active transport then they will compete for it when they are both present in the solution. With a limited amount of carrier available both have their rates of absorption reduced. The L-glucose however is clearly absorbed preferentially to the L-galactose by the carrier and consequently the reduction in the rate of uptake is less marked for L-glucose.

(e) D-glucose is an optical isomer of L-glucose. Structurally the molecular configurations have only one minor difference. This however is enough for the highly specific carrier molecules to distinguish between the two isomers. Since L-glucose is the form found in living organisms, the carrier system has evolved to absorb this type and not the D-glucose.

Example 11 (Time allowance 2 minutes)
The graph below shows the changes in various pressures which occur when water enters or leaves a plant cell.

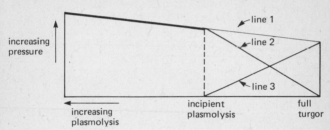

Study the graph and then state the line number which refers to:

(a)   wall pressure (WP)   (b)   diffusion pressure
(c)   osmotic pressure (OP)   deficit (DPD)
(d)   water potential (ψ)   (e)   turgor pressure (TP)

Answer
(a)   Line 3   The wall pressure (the pressure the cell wall exerts on the cell contents) is equal to 0 while the cell is plasmolysed and increases to a maximum when the cell is fully turgid (i.e. line 3).
(b)   Line 2   The DPD is the net tendency of a cell to draw in water from its surrounding solution. This influx of water is opposed by the wall pressure and so decreases as the wall pressure increases (i.e. line 2).
(c)   Line 1   The OP is in proportion to the solute concentration of the cell. As water is slowly drawn into a cell its solute contents become slowly diluted and so its OP gradually falls (i.e. line 1).
(d)   Line 2   Water potential is an alternative term for DPD.

(e) Line 3 The turgor pressure is the pressure the cell contents exert on the cell wall. It is equal in magnitude (although opposite in direction) to the wall pressure. It therefore follows the same line as the wall pressure (i.e. line 3).

Example 12 (Time allowance 8 minutes)
(a) If a plant cell has a turgor pressure of 3 atmospheres and an osmotic potential (OP) of 6 atmospheres, what is the diffusion pressure deficit (DPD) of the cell?
(b) What would happen to a fully turgid plant cell with a diffusion pressure deficit of 5 atmospheres if it were placed in a sucrose solution with an osmotic potential of 6 atmospheres?
(c) If the osmotic potential of 0·1 mole per litre sucrose is equal to 2·6 atmospheres what is the osmotic potential of a 0·5 mole per litre sucrose solution?
(d) Does a 1 mole per litre solution of sodium chloride have the same osmotic potential as a 1 mole per litre solution of sucrose? Explain your answer.

Answer
(a) The turgor pressure of a cell is equal in magnitude, although opposite in direction, to the wall pressure. The wall pressure (WP) is therefore equal to 3 atmospheres.
Substituting in the equation DPD = OP − WP
$$DPD = 6 - 3$$
Therefore DPD = 3 atmospheres
(b) The external pressure drawing water from the cell is equal to 6 atmospheres (the OP of the sucrose solution). This is slightly greater than the DPD of the cell (5 atmospheres) which is drawing water in. Water is therefore slowly drawn out and the cell becomes slightly plasmolysed, i.e. the protoplast is drawn away from the cell wall.
(c) The osmotic potential of a dilute solution is directly proportional to the molar concentration for non-ionizing solutions. In this case therefore the OP of 0·5 molar sucrose solution will be five times greater than the OP of a 0·1 molar sucrose solution.
i.e. 2·6 × 5 = 13 atmospheres
(d) No—the direct relationship between osmotic potential and molar concentration applies only to non-ionizing solutions. In effect it is the number of particles that determines the osmotic potential. Sodium chloride ionizes into sodium and chloride ions thus doubling the number of particles and giving it approximately twice the OP of an equivalent molar solution of sucrose.

Example 13 (Time allowance 1 minute)

The sap of a plant cell has an osmotic pressure of 12 atmospheres and there is a wall pressure of 4 atmospheres. When this cell is placed in a solution with an osmotic pressure of 5 atmospheres the force causing water to enter the cell is

A 9 atmospheres.
B 8 atmospheres.
C 5 atmospheres.
D 3 atmospheres.

Answer D

The equation for this question is DPD = OP − WP. Here the DPD is 'the force causing water to enter the cell'. However the calculation is not so straightforward as it would be if the cell were placed in pure water. If this were the case then the DPD would be equal to the cell sap osmotic pressure (12 atmospheres) minus the wall pressure (4 atmospheres), i.e. 8 atmospheres. However the solution into which the cell is placed has its own osmotic pressure of 5 atmospheres and hence the DPD is reduced by this amount, compared to a solution of pure water. The true force by which water enters the cell is hence only (8 − 5) which is 3 atmospheres.

The calculation should be based on:

DPD (external) + DPD (internal) = OP − WP
i.e.   5   + DPD (internal) = 12 − 4
Therefore DPD (internal)   = 3 atmospheres

The correct response is therefore D.

Chapter 6   Nutrition

Food chains, food webs and energy levels

Example 1 (Time allowance 10 minutes)
(a)   Study the following sentences and then relate each one to one of the five terms A, B, C, D and E.
1   A natural community of plants and animals.
2   The study of the interrelationships between living organisms and their environment.
3   The wise management and use of natural resources.
4   That part of the earth and its atmosphere inhabited by living things.
5   A naturally occurring group of organisms inhabiting a common environment.
A   ecology
B   conservation
C   community
D   ecosystem
E   biosphere
(b)   A simple food web of five organisms A—E is shown below.

If organism C were suddenly to be removed from the food web, how would organisms A, D and E be affected? Explain your answers.

Answer
(a)   1 D   2 A   3 B   4 E   5 C
(b)   A   The direction of the arrow from organism A to organism C indicates that energy is flowing from A to C. This effectively means that A is eaten by C. If C were to suddenly disappear, one means by which A is destroyed would be removed and the population of A would increase in size. The increase would not be indefinite because A is also eaten by organism B and in the event of the population of A increasing, B would have a larger food supply and its population too would increase. More A would be consumed until a new equilibrium between the two populations was established.
D   The arrow indicates that C is eaten by D. The sudden removal of C would therefore remove one source of food for D whose population would decrease. To what extent it would decrease

42

would depend upon how much the shortfall in food could be compensated for by consuming more of organism B. The organism however is unlikely to disappear altogether as it has B as an alternative food source.

E   Assuming that there is no other food source for organism E than organism C, then the complete removal of C would result in the extinction of E in due course. The time lag between the removal of C and the extinction of E would depend on the internal and external food stores of E.

Example 2 (Time allowance 1 minute)
Which of the following components of an ecosystem has the greatest biomass?
A   primary producers
B   primary consumers
C   secondary consumers
D   tertiary consumers
E   decomposers

Answer A
At each stage of a food chain or web some energy is lost. The biomass is therefore always greater at the beginning of any food chain. The primary producers are always at the start of any food chain and therefore have the greatest biomass.

Example 3 (Time allowance 7 minutes)
The distribution of four species of organisms at different depths in a pond was investigated and the data presented graphically as shown below.

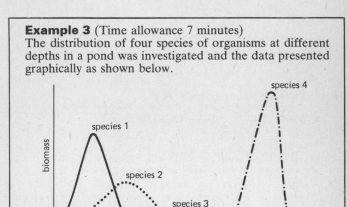

43

> (a) (i) Which species is most likely to be the main primary producer?
> (ii) Give two reasons for your choice.
> (b) (i) Which species is most likely to be a secondary consumer?
> (ii) Give a reason for your choice.

Answer

(a) (i) Species 1.

(ii) 1 Primary producers are photosynthetic and therefore are most likely to occur near the surface of a pond where light intensity is greatest. Species 1 has its maximum biomass nearer the surface than the remaining species.

2 Species 4 is found exclusively at depths of two metres or greater. Light intensity at this depth could be very low and photosynthesis impossible therefore species 4 is not the main primary producer. Although the remaining three species could be primary producers, species 1 has a greater biomass than the other two and must therefore be the main primary producer.

(b) (i) Species 4.

(ii) It has been shown in (a) that species 1 is the main primary producer and this can therefore be discounted. A primary consumer feeds on the primary producers and must therefore be found at the same depth. Species 2 and 3 could therefore each be primary consumers as they occur at the same depth as species 1. Species 4 however cannot be a primary consumer as it is never found at the same depth as the primary producer (species 1). It has been established in (a) that species 4 cannot itself be a primary producer and it must therefore be a secondary consumer at least. It presumably feeds on species 3 at around 2·0 m depth.

Photosynthetic autotrophs

Example 4 (Time allowance 10 minutes)
(a) State five conditions necessary for photosynthesis to occur.
(b) Briefly describe a simple experiment to show that gaseous oxygen is evolved as a product of photosynthesis.

Answer

(a) 1 A suitable temperature (i.e. in the range 5°-30°C)
2 A supply of water
3 A supply of carbon dioxide
4 Sunlight   5 The presence of chlorophyll

(b) Some pieces of a water plant such as *Elodea* should be placed under an inverted funnel in a beaker full of dilute sodium hydrogen carbonate. A test tube filled with dilute sodium hydrogen carbonate solution should be inverted over the stem of the funnel as shown in the diagram which follows.

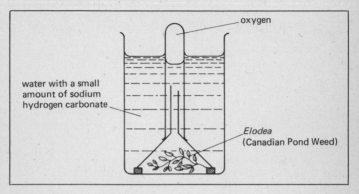

The sodium hydrogen carbonate provides a source of carbon dioxide to allow *Elodea* to photosynthesize. The apparatus is exposed to bright light. The bubbles of gas arising from the weed collect in the test tube which when full of gas is removed from the apparatus. If a glowing splint is inserted into the test tube of gas it will relight if the gas present is oxygen.

Example 5 (Time allowance 12 minutes)
The graph below shows the rate of photosynthesis in two species of plant at different light intensities.

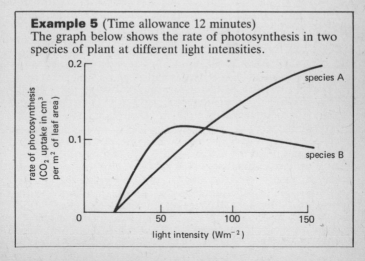

(a) Which species shows the best adaptation to shade conditions?

(b) Give reasons for your answer.

(c) Apart from intensity, state one other way in which light in a shady area differs from that in a sunny area.

(d) Many species of plant that grow in the shade have low rates of respiration. What is the possible advantage of this?

(e) From the graph it appears that photosynthesis does not begin until a light intensity of 20 W m^{-2}. Explain why this is so.

Answer

(a) Species B.

(b) The rate of photosynthesis for species B is greater than species A at lower light intensities (i.e. 20-75 W m^{-2}). This is the range of light intensities more likely to be encountered in shade conditions.

The rate of photosynthesis for species B begins to decrease at light intensities above 75 W m^{-2} whereas that of species A continues to increase. Species A clearly photosynthesizes well at high light intensities whereas species B actually shows a fall in the rate of photosynthesis at these light levels.

(c) In a shady area much of the light has been filtered through, and/or reflected from, other vegetation. This light will therefore be of an overall lower wavelength than light in a sunny area.

(d) Respiration utilizes the products of photosynthesis (oxygen and glucose). Plants growing in the shade have a relatively low rate of photosynthesis due to the lower light intensity with consequent smaller yields of oxygen and glucose. If the plant is to retain sufficient glucose to synthesize into other material, such as protein, in order to grow, produce flowers etc., it must minimize the amount of glucose oxidized in respiration. The rate of respiration is therefore reduced.

(e) The rate of photosynthesis is measured as the carbon dioxide uptake in cm^3 per m^2 of leaf area. Although photosynthesis does occur at light intensities below 20 W m^{-2} the carbon dioxide taken in is obtained from the carbon dioxide produced during respiration. There is therefore no actual uptake of carbon dioxide at these light intensities. Above light intensities of 20 W m^{-2} the rate of carbon dioxide uptake by photosynthesis exceeds its production in respiration and a net uptake of carbon dioxide occurs.

Example 6 (Time allowance 5 minutes)
(a) What is meant by the 'concept of limiting factors'?
(b) If light were limiting the rate of photosynthesis in an aquatic plant in a laboratory experiment, which of the following would increase the rate of photosynthesis fourfold?
1 Increasing the carbon dioxide concentration four times.
2 Increasing the temperature by 20°C.
3 Increasing the temperature four times.
4 Halving the distance between the plant and the light source.
5 Reducing the distance between the plant and the light source to one quarter of its original distance.

Answer
(a) Any process requiring a number of different conditions for its success will proceed at a rate determined by the factor in shortest supply and that factor alone, i.e. the factor nearest its minimum value limits the rate of the reaction.
(b) 4 If light is limiting the reaction, only a change in light intensity can increase the rate. Options 1, 2 and 3 may therefore be immediately discounted. The light intensity is inversely proportional to the square of the distance from the source, i.e. if the distance between the plant and the light source was reduced by one quarter, the rate of photosynthesis would be increased by 4^2 (=16 times). This is more than required by the example and so option 5 may be discounted. Halving the distance increases the rate of photosynthesis by 2^2 (=4 times). This makes option 4 the correct response.

Example 7 (Time allowance 6 minutes)
Complete the blanks in the following passage concerning the structure and function of a leaf.
A leaf comprises the stalk called the . . . .(A). . . . . and a flat blade called the . . . .(B). . . . . It is bounded on both upper and lower surfaces by a waterproof . . . .(C). . . . . which covers a single-celled layer called the . . . .(D). . . . . In this layer are pores called . . . .(E). . . . . each of which is bounded by a pair of . . . .(F). . . . . The opening and closing of these pores is partly controlled as follows: In daylight the rate of . . . .(G). . . . . increases and so . . . .(H). . . . . is rapidly absorbed by the plant. The pH of the cells surrounding the pores therefore

. . . .(I). . . . . . . This causes. . . . .
. . . .(J). . . . . in the cells to be converted to . . . .
. . . .(K). . . . . and the . . . .
(L). . . . . . . pressure of the cells rises. Water is drawn
into the cells causing a rise in . . . .
(M). . . . . . . pressure. The variation in thickness be-
tween the inner and outer walls of the cells causes the pores
to . . . .(N). . . . . . .

Answer

A   petiole   B   lamina   C   cuticle   D   epidermis
E   stomata   F   guard cells   G   photosynthesis
H   carbon dioxide   I   rises/increases   J   starch
K   glucose (sugar)   L   osmotic   M   turgor   N   open

Example 8 (Time allowance 2 minutes)
Name four factors which would affect the rate of transpiration in a healthy green plant.

Answer

Choose from:

1 Carbon dioxide concentration	4 Humidity	7 Atmospheric pressure
2 Temperature	5 Wind/ air currents	8 Size of stomatal aperture
3 Light intensity	6 Water supply	

Example 9 (Time allowance 4 minutes)
  (a) What is meant by the term 'action spectrum' in
       relation to photosynthesis?
  (b) Give the colour and approximate wavelength of the
       two bands of light that give the maximum rate of
       photosynthesis.
  (c) Give the name and colour of four pigments normally
       found in a chloroplast.

Answer

(a) It is the relative amount of photosynthesis that takes place at
different wavelengths of light.
(b) (i)   Blue light in the range 400–450 nm (peak at 430nm)
     (ii) Red light in the range 600–700 nm (peak at 660nm)
(c) (i)   Chlorophyll a—blue/green
     (ii) Chlorophyll b—yellow/green
     (iii) Xanthophyll—yellow (iv) Carotene—yellow/orange
The above four are the best answers, although phaeophytin
(grey/brown) could possibly be substituted. However it is thought
that this is a breakdown product.

Example 10 (Time allowance 6 minutes)
Photosynthesis has two distinct stages, the light stage and the dark stage.
(a) The diagram below is a simplified sequence of the dark stage:

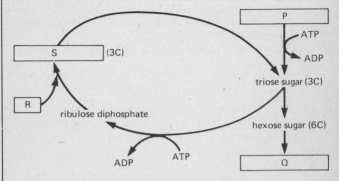

(i) Name the substances P, Q, R and S.
(ii) How many carbon atoms has ribulose diphosphate?
(iii) From which intermediate may amino acids be synthesized?
(iv) Which two intermediates are needed to form fats?
(b) What are the products of the light stage?

Answer
(a) (i) P   nicotinamide adenine dinucleotide phosphate (reduced) $NADPH_2$ Q   starch   R   carbon dioxide   S   phosphoglyceric acid (PGA)
(ii) Five (iii) Phosphoglyceric acid
(iv) The triose sugar (3C) which can be converted to glycerol and phosphoglyceric acid which can be converted to fatty acids. Fatty acids + glycerol = fat
(b) Oxygen (from the photolysis of water)
$NADPH_2$ (reduction of NADP is achieved using H^+ from the photolysis of water)
ATP (produced from ADP by photophosphorylation)

Example 11 (Time allowance 5 minutes)
State two functions in the green plant of the following minerals: (i) magnesium (ii) nitrogen (iii) phosphorus (iv) potassium.

Answer
(i) 1 It is a constituent of the chlorophyll molecule and therefore essential for photosynthesis and growth.
2 It is an enzyme activator in several of the enzymatic steps of glucose degradation during respiration.
(ii) 1 It is a major constituent of all proteins and therefore essential for growth.
2 It is a constituent of chlorophyll and plant hormones—again essential for growth.
(iii) 1 It is found in some proteins and therefore needed for growth.
2 It is a constituent of adenine triphosphate, the energy-storage material essential in photosynthesis and respiration. It is also found in DNA and RNA.
(iv) 1 It is required for normal cell division.
2 It is involved in the synthesis and translocation of carbohydrate and in the synthesis of protein in meristematic cells.

Holozoic heterotrophs

Example 12 (Time allowance 10 minutes)
The following table represents the results of an experiment in which the stomach contents of 20 individuals were examined at regular intervals. In each case the individuals were given 100g of a particular food and by removing some of the stomach contents through a narrow tube, the length of time the food-stuff remained in the stomach was ascertained. At the same time the amount of acid present in the stomach was calculated by titration with suitable alkali.

Food eaten	Length of time food remained in stomach (minutes)	Amount of acid present (Arbitrary units)
Bread	140	11
Rice (boiled)	130	12
Sugar (sucrose)	105	8
Cabbage	110	8
Apple	100	7
Chicken	200	15
Pork	200	15
Beef	175	16
Fish	160	15

(a) Which **type** of food remains in the stomach for the longest period of time?

(b) Explain why this type of food should remain in the stomach longer than other types.

(c) What correlation is there between the acid level and the length of time any particular food remains in the stomach?

(d) Explain this correlation.

(e) If the subject had been given green peas as one of the experimental foods, estimate the time it would have remained in the stomach and the acid level present. Give reasons.

Answer

(a) Protein.

(b) The stomach contains pepsin, an enzyme that breaks down proteins into polypeptides. The hydrochloric acid produced by the stomach wall provides a pH of 2–3, the optimum for pepsin action. Conditions are therefore ideal for protein digestion which takes some time to complete. The protein therefore remains in the stomach for the longest period of time.

(c) In general the longer the food remains in the stomach the greater is the acid level.

(d) Firstly the hydrochloric acid is needed to convert the inactive pepsinogen produced by the gastric glands into active pepsin; it also provides the optimum pH for pepsin to work. As protein remains longest in the stomach it is hence necessary for hydrochloric acid production to continue throughout this period. Secondly the presence of food in the stomach stimulates the stomach lining to produce gastric juice and the hormone gastrin. This hormone circulates in the blood and causes the oxyntic cells of the stomach wall to produce hydrochloric acid. So the longer food is in the stomach the more gastrin and acid is produced.

(e) Although not entirely accurate, it is a fair generalization to say that the length of time a food remains in the stomach reflects its protein content and the acid level is correlated to the time it remains in the stomach (answer c). Peas have a high protein content for plant material, although less than a similar weight of meat or fish. Their protein value probably lies somewhere between bread and fish. They would probably therefore remain in the stomach for an intermediate time and have an intermediate acid level, i.e. about 150 minutes and 13 units respectively.

Example 13 (Time allowance 1 minute)
Which of the following best represents the enzyme composition of pancreatic juice?
A   Amylase, peptidase, trypsinogen, rennin
B   Amylase, pepsin, trypsinogen, maltase
C   Lipase, amylase, pepsin, maltase
D   Lipase, amylase, trypsinogen, peptidase
E   Peptidase, amylase, pepsin, rennin

Answer   D
All other options may be discounted because they contain one or more of the following substances which are produced elsewhere as indicated: rennin and pepsin (gastric glands of the stomach wall); maltase (the wall of the small intestine).

Example 14 (Time allowance 1 minute)
The main function of the bile salts is to:
A   neutralize acid chyme.
B   excrete breakdown products of haemoglobin.
C   hydrolyse fats into fatty acids and glycerol.
D   emulsify fats.
E   stimulate the contraction of the gall bladder.

Answer   D
Options A, B and D are all functions of the constituents of bile juice, however A is carried out by mineral salts such as sodium bicarbonate and B is carried out by the bile pigments biliverdin and bilirubin. Only D is carried out by the bile salts. Option C is performed by lipase from the pancreas (there are no enzymes in bile juice). Option E is a function of cholecystokinin-pancreozymin (CCK-PZ).

Example 15 (Time allowance 1 minute)
Many herbivorous mammals have a specialized region of the alimentary canal that contains micro-organisms. Which of the following best describes the function of these micro-organisms?
A   They produce vital vitamins essential to the herbivore.
B   They produce an enzyme that converts cellulose into starch.
C   They produce an enzyme that hydrolyses starch into maltose.
D   They produce nitrogenous waste products which are beneficial to the herbivore.
E   They produce an enzyme which hydrolyses cellulose.

Answer E
The micro-organisms do not directly produce vitamins (option A). Although they produce an enzyme that acts on cellulose, it does not convert it to starch (option B). The enzyme hydrolyses cellulose not starch (option C) and while micro-organisms may be digested later to provide protein for the mammal, any nitrogenous waste products are produced in insignificant amounts, thus eliminating option D.

Example 16 (Time allowance 8 minutes)
(a) List the differences between the dentition of a herbivore such as a sheep and those of a carnivore such as a dog.
(b) Why is lateral jaw movement found in herbivorous mammals but not in carnivorous ones?
(c) State one general feature that is common to the habitats of carnivorous plants.
(d) Correlate this feature to the carnivorous nature of these plants.

Answer

(a)

Herbivore	Carnivore
1 Horny pad—no upper incisors	Upper incisors present and pointed
2 Lower incisors chisel-shaped	Lower incisors pointed
3 No canines	Long pointed canines
4 Diastema (gap between incisors and premolars) present	Diastema absent
5 Flat ridged premolars and molars	Sharp pointed premolars and molars
6 Teeth have open roots and grow continuously	Teeth have closed roots and do not grow continuously

(b) Herbivorous mammals need to grind the vegetation with their ridged premolars and molars in order to break the cellulose cell wall of the plant cells. Effective grinding requires lateral jaw action as well as backwards and forwards movement. Such lateral jaw motion involves a less rigid articulation of the lower jaw and consequently the jaw is more easily dislocated. This would be a major disadvantage to a carnivore leading to the possible escape of prey which struggled in the mouth. Carnivores as a result lack lateral jaw movement.
(c) The substrate lacks some essential mineral, usually nitrogen.
(d) The absence of the mineral means that the plant needs to capture small animals and to digest them as a means of obtaining the missing mineral.

Example 17 (Time allowance 10 minutes)
(a) Give **one** example of (i) an endoparasite, (ii) an ectoparasite.
(b) Explain why parasites produce large numbers of offspring.
(c) Differentiate between an obligate and facultative parasite.
(d) (i) List **three** ways in which parasites and saprophytes are similar.
(ii) List **three** ways in which parasites and saprophytes differ.
(e) State **four** advantages of saprophytes to man.

Answer
(a) (i) Choose from: *Plasmodium* (malarial parasite); *Taenia* (tapeworm); *Fasciola* (liverfluke).
(ii) Choose from: *Pulex* (flea); *Argas* (tick); *Sarcoptes* (mite).

(b) The parasite requires a new host to survive because the present host cannot live indefinitely. Most parasites are highly host-specific and must find a suitable member of a single species. This in itself is an unlikely event and the production of large numbers of offspring is essential to increase the likelihood of success. In addition, a period outside the host is often necessary and this involves the loss of many individuals. A large number of offspring is essential to compensate for this.

(c) An obligate parasite is obliged to live parasitically and cannot survive in any other way, whereas a facultative parasite normally lives parasitically but may continue to feed saprophytically on the host once it has died.

(d) (i) Choose from:
1 show heterotrophic nutrition
2 absorb simple food substances
3 have simple digestive systems (if present)
4 show sexual and asexual phases of reproduction, often with resistant stages
5 produce large numbers of offspring

(ii) Choose from:

Parasites	Saprophytes
1 Energy is derived from living organisms	Energy is derived from dead organisms
2 Very specific to their hosts	Use a wide variety of food sources
3 Nutritionally very specialized and highly adapted	Simple nutritional methods
4 Most plant and animal groups include parasites	Almost totally limited to bacteria and fungi

5 Mostly aerobic   Aerobic and anaerobic
6 Often many stages to the   Usually a single adult stage
   life cycle

(e) Choose from:
    1 production of antibiotics, e.g. penicillin
    2 recycling of nutrients, e.g. carbon, nitrogen—essential
       for crop production
    3 brewing and baking using yeast (*Saccharomyces*)
    4 food source, e.g. mushrooms
    5 manufacture of yoghurt and cheese
    6 industrial processes, e.g. tanning of leather, production
       of vitamins
    7 decomposition of sewage

Example 18 (Time allowance 7 minutes)
A lichen is a symbiotic association between two types of organisms.
(a)   Name the two types of organism.
(b)   Describe the benefits each derives from the other.
(c)   What is a mycorrhizal association?
(d)   Using an example to illustrate your answer explain
       what is meant by commensalism.

Answer
(a) An alga and a fungus.
(b) The alga produces oxygen and carbohydrate during photosynthesis which it gives to the fungus for use in respiration. The fungus provides water and carbon dioxide from its respiration for use by the alga in photosynthesis. In addition the fungus may absorb some essential minerals which it passes on to the alga. It also surrounds the alga protecting it from desiccation and thereby allowing it to survive in arid areas which it would not otherwise be able to colonize.
(c) It is the association of certain soil fungi with plant roots. The affected roots are shorter, stouter and more branched forming a coral-like mass. The fungal hyphae form around the root tips and penetrate the cortical cells. The fungus aids the plant with the uptake of certain minerals such as nitrogen and phosphorus and in return is given carbohydrate.
(d) In commensalism the commensal always gains but the host neither gains nor loses. The hermit crab *Pagurus bernhardus* is host to a colonial hydroid *Hydractinia echinata* which lives on its shell. The hydroid benefits from consuming food particles which are released when the crab devours its food and it is taken by the crab into regions it could not visit by itself. The crab neither benefits, nor loses by the relationship.

Example 19 (Time allowance 1 minute)
When legumes are grown on sterile soil they do not develop fully. The addition of certain living, nitrogen-fixing bacteria to the soil improves growth. These bacteria can be isolated from nodules that form on the roots and can be shown to utilize carbohydrates formed from the legumes. The relationship between the organisms is:

A commensalism   B symbiosis   C parasitism
D saprophytism

Answer B

A synopsis of the information is 'in the absence of bacteria legume growth is poor, but in their presence legumes grow well. The bacteria utilize legume carbohydrate'. A is incorrect because both, not just one, organisms benefit from the relationship. C is incorrect for the same reason plus the fact that neither organism is harmed, and D can be eliminated because a saprophyte lives off dead material and the legume is clearly alive if its growth is improving. The possibility of the legume living saprophytically on the bacteria may be discounted as all legumes are autotrophic. Both organisms derive benefit from the relationship which is therefore symbiotic.

Example 20 (Time allowance 1 minute)
The type of nutrition exhibited by a scavenger is best described as:

A   parasitic.
B   saprophytic.
C   autotrophic.
D   symbiotic.
E   holozoic.

Answer E

To be a parasite (option A) the organism upon which the scavenger feeds would need to be living. Scavengers are normally animals feeding off dead remains. In addition the relationship between the scavenger and its food is not as permanent as that between the parasite and its host. The word 'saprophyte' (option B) accurately refers to plants only, although it is often used more loosely. Scavengers are never plants. In any case saprophytes live in close association with their food which is **not** the case with scavengers. Scavengers are always animals and so cannot be autotrophic (option C). To be symbiotic (option D) both organisms must benefit, and to do so must be living. As the scavengers' food is dead this option can be discounted. Scavengers are holozoic (option E).

Chapter 7 Respiration
Cellular respiration

Example 1 (Time allowance 1 minute)
In the hydrogen carrier system of aerobic respiration, which of the following is the last stage?
A formation of ATP
B reduction of oxygen
C production of carbon dioxide
D reduction of cytochrome
E production of H^+

Answer B
By acting as the final acceptor of hydrogen ions, it is the oxygen that 'drives' the hydrogen carrier system. In the process the oxygen is reduced to water, one of the products of respiration. Although the formation of ATP (option A) does occur in the hydrogen carrier system it is not the last stage. The production of carbon dioxide (option C) and the production of H^+ (option E) occur mainly in the Krebs' cycle but never in the hydrogen carrier system. Reduction of cytochrome (option D) is the penultimate stage of the hydrogen carrier system.

Example 2 (Time allowance 4 minutes)
(a) Why, during the early stages of glycolysis, is glucose phosphorylated?
(b) State two roles played by ATP in this process of phosphorylation.
(c) When glucose is split into two triose sugars it may be converted to pyruvate. What two other possible metabolic fates are there for this triose sugar?

Answer
(a) To make the glucose molecule more reactive and so allow it to be split more easily into triose sugar.
(b) (i) It provides the phosphate groups.
(ii) It provides the energy to activate the glucose.
(c) (i) It may be converted into glycerol and ultimately into fat.
(ii) It may be converted into certain amino acids and ultimately into protein.

Example 3 (Time allowance 5 minutes)
The following is a simplified scheme of a metabolic process:

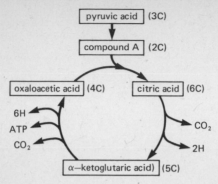

(a) Name the cycle shown.   (b)   Identify compound A.
(c) Give general names for the **two** biochemical processes involved in the conversion of α-ketoglutaric acid to oxaloacetic acid.
(d) What is the primary acceptor of the hydrogens produced in this cycle?
(e) What compound is ultimately produced from the transfer of these hydrogens from carrier to carrier?

Answer
(a) Krebs' (Citric acid) (Tricarboxylic acid) cycle.
(b) A = Acetyl co-enzyme A
(c) Decarboxylation (i.e. removal of carbon dioxide)
    Dehydrogenation (i.e. removal of hydrogen)
(d) Nicotinamide adenine dinucleotide (NAD)
(e) Adenosine triphosphate (ATP)

Example 4 (Time allowance 10 minutes)
(a) What are the end products of anaerobic respiration in
    (i) animals? (ii) yeast?
(b) State **four** differences between aerobic and anaerobic respiration.
(c) (i) Where in a mammal does anaerobic respiration most commonly occur?
    (ii) What advantage is there in anaerobic respiration occurring here?

Answer
(a) (i) Lactic acid (CH_3CHOH $COOH$) **(ii)** Ethanol (CH_3CH_2OH)
(b) Choose any four from the following table:

Aerobic respiration	Anaerobic respiration
Free molecular oxygen is required	Free molecular oxygen is **not** required
More efficient energy release (about 55% of that available)	Less efficient release of energy (around 2% of that available)
Complete breakdown of glucose occurs	Glucose breakdown is not complete
The end products are less toxic	End products are more toxic
The process is similar in plants and animals	The process is different in plants (alcohol is the end product) and animals (lactic acid is the end product)
Krebs' cycle occurs	Krebs' cycle does not occur
Process occurs in cytoplasm and mitochondria	Process takes place in cytoplasm only
More complex pathway	Less complex pathway
Glycolysis is slower	Glycolysis occurs faster

(c) (i) In skeletal muscle tissue.
(ii) Skeletal muscle is used largely in locomotory activities. If the mammal should be involved in some vital activity such as escaping a predator or catching prey it is often necessary for the locomotory activity to be intensive and/or prolonged. In these circumstances the blood may be unable to supply oxygen at a rate adequate to meet the energy needs of the organism. Anaerobic respiration however allows enough energy to be released to sustain the activity for longer.

Example 5 (Time allowance 20 minutes)
(a) What is meant by 'respiratory quotient'?
(b) Two plant seedlings X and Y of different species had the following respiratory quotients during their early development.

Days from the start of germination	X	Y
1	0·61	0·65
5	0·41	0·91
9	0·71	0·99
13	0·70	1·02

From the results discuss the possible nature of the respiratory substrates being used by X and Y.

Answer
(a) The respiratory quotient is the ratio of the volume of carbon dioxide expired to the volume of oxygen consumed during the same period.

$$RQ = \frac{\text{carbon dioxide produced}}{\text{oxygen used}}$$

(b) Typical RQ values for the breakdown of common chemicals found in seeds and seedlings are:
Carbohydrate = 1·0
Fat   = 0·7
Protein   = 0·8–0·9
In the initial stages of germination species X does not have an RQ corresponding to any of the above chemicals. Fat is the most common storage material in seeds because it has a relatively small mass for a given amount of stored energy. In the early stages of germination this fat is not all respired directly, but some is converted to carbohydrate according to the overall equation:

$$C_{16}H_{32}O_2 + 11O_2 \rightarrow C_{12}H_{22}O_{11} + 4CO_2 + 5H_2O$$

$$RQ \therefore = \frac{CO_2 \text{ evolved}}{O_2 \text{ taken up}} = \frac{11}{4} = 0.35$$

It could be therefore that on day 1 species X is respiring some fat (RQ = 0·7) but a little is being converted to carbohydrate (RQ = 0·35), the result is an RQ of 0·61. As germination proceeds a greater proportion of the fat is converted to carbohydrate and the RQ consequently falls to 0·41 on day 5. By day 9 the conversion of fat to carbohydrate has ceased and fat is directly respired to give an RQ of 0·71. This situation continues until day 13. An alternative explanation could be that from day 9 to day 13 some fat is being converted to carbohydrate (RQ = 0·35) and this carbohydrate (or that produced in photosynthesis) is being respired (RQ = 1·00). The result of these two processes occurring concurrently is an RQ of around 0·7.
For species Y it appears that on day 1 mostly fat is being respired (RQ = 0·7) with a little being converted to carbohydrate (RQ = 0·35) giving a resultant RQ of 0·65. Thereafter the carbohydrate produced (or stored) is respired (RQ = 1·0) along with some fat (RQ = 0·7) giving an overall RQ of 0·91. As photosynthesis begins, more carbohydrate is produced and the store of fat is finally used up. Carbohydrate is respired, resulting in an RQ of about 1·0.

Gaseous exchange

Example 6 (Time allowance 30 minutes)
Below is a diagram of a simple respirometer, which is used to measure the volume of oxygen taken up by organisms.

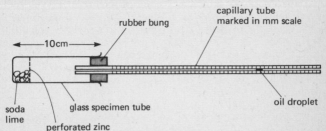

(a) Suggest an organism which could be used in the apparatus.
(b) Briefly describe how the apparatus works and how you would use it.
(c) What control could be set up and how might it be used to modify the results obtained?
(d) List **four** sources of error that could arise when using the apparatus.
(e) State **four** factors that could alter the rate of energy uptake by the organisms in the apparatus.
(f) If, during one experiment, the oil droplet moved 50mm in 10 minutes and the total mass of the organisms in the glass specimen tube was 4g, calculate the volume of oxygen taken up per hour, per gram of organism. N.B. The capillary tube has a uniform bore of 1·0mm.

Answer
(a) Choose from : locust, beetle, housefly, earthworm, wood-louse, maggot, bee, germinating pea seeds.
(b) The apparatus should be set up as above. A sample of one or more of the appropriate organisms should be weighed and put in the specimen tube. By inserting the bung further into the tube the oil droplet should be positioned as far away from the specimen as possible. The position of the droplet in the capillary tube should be recorded from the scale marked on it and a stop clock started. As the animal(s) respire according to the equation:

$$6O_2 + C_6H_{12}O_6 \rightarrow 6CO_2 + 6H_2O + energy$$

the volume of carbon dioxide produced is equal to the volume of

61

oxygen taken up. The carbon dioxide however is rapidly absorbed by the soda lime and its volume therefore becomes negligible. The volume of water is also negligible as this too is absorbed by the soda lime. The only measurable volume change is therefore due to the reduction in oxygen volume as it is absorbed by the animals. This reduction in volume causes a reduction in pressure within the specimen tube. Atmospheric pressure now exceeds the internal pressure and the oil droplet is forced towards the specimen tube. The time taken for the droplet to move a set distance (e.g. 15cm) should be recorded and the process repeated a few times to allow an average time to be obtained. If the distance moved is 'h' then the volume of oxygen consumed may be calculated from the formula $\pi r^2 h$. 'r' is the radius of the capillary tube and may be calculated by finding the internal bore diameter (using a travelling microscope) and dividing by two. The volume consumed per minute is calculated by dividing the total volume by the number of minutes taken for the droplet to move the distance 'h'. This figure should then be divided by the weight in grams of the animals to give a final figure of oxygen consumed/minute/g.

(c) A second set of apparatus, as identical as possible to the experimental one, should be set up as a control, in exactly the same way as described in (b), except that the animals should be excluded. This control should be placed as close as possible to the experimental one in order that the environmental fluctuations affect both equally. If there is no movement of the droplet in the control apparatus the results of the experimental apparatus require no alteration. If, however, the changes in atmospheric pressure or temperature, cause the droplet in the control tube to move, the distance moved should be recorded. This distance should be added to the distance 'h' measured in the experimental tube if the droplet moves away from the specimen tube in the control, or subtracted if it moves towards the specimen tube. In this way the control acts as a thermobarometer.

(d) Choose from:
   (i) apparatus not air tight
   (ii) temperature changes during the experiment (will alter pressure)
   (iii) pressure changes during the experiment
   (iv) soda lime may be exhausted (i.e. has ceased to absorb more carbon dioxide)
   (v) capillary tube may not be horizontal (gravity may move droplet)

(e) Choose from:
   (i) external temperature fluctuations
   (ii) age of the animals used
   (iii) activity of the animals (i.e. resting or moving excitedly)
   (iv) amount of oxygen available in the specimen tube

(v)   light intensity (may cause increase/decrease in activity depending on the animal examples used)

(f)   The movement of the oil droplet means that the organisms have effectively absorbed a cylinder of oxygen of height 50mm.

Volume of cylinder $= \pi r^2 h$

Here h — 50mm

$$r = \frac{1\cdot0mm}{2} \text{ (diameter of cylinder)} = 0\cdot5mm$$

Therefore total amount of oxygen absorbed $= \pi \times (0\cdot5)^2 \times 50$

This for a period of 10 minutes, therefore the amount of oxygen taken up in 1 hour $= \pi \times (0\cdot5)^2 \times 50 \times 6$.

This for 4g of organisms, therefore the amount of oxygen taken up per hour, per gram of organism

$$= \frac{\pi \times (0\cdot5)^2 \times 50 \times 6}{4} = \mathbf{58\cdot9mm^3\ O_2/hour/g}$$

Example 7 (Time allowance 25 minutes)

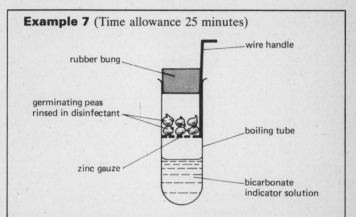

Above is the diagram of a simple laboratory experiment, study it carefully and then answer the following questions:
(a)   What is the experiment designed to show?
(b)   Describe a suitable control experiment.
(c)   What changes would you expect in your two experiments at the end of 2 hours? Explain your answer.
(d)   What adaptations would be necessary if the peas were replaced by an aquatic plant such as *Elodea*?

Answer
(a)   The experiment shows that carbon dioxide is produced by germinating peas. (Although it is most likely that this carbon dioxide is produced during respiration, the experiment cannot determine the exact source of the carbon dioxide.)

(b) Any control experiments should differ from the original experiment in a single respect only, namely the factor being investigated. This experiment attempts to show that the germinating peas produce carbon dioxide, rather than the bicarbonate indicator, test tube, gauze or rubber bung. One control experiment would be to set up a second set of apparatus, similar to the first and under the same environmental conditions, except that the peas should be omitted. In some respects a better control is to take a sample of peas equal in weight to those in the first experiment. These should be boiled for about 10 minutes in a disinfectant (e.g. formaldehyde) solution. This both kills the peas and any micro-organisms on them, by denaturing their enzymes. These dead peas should then be used in the control experiment which again should be carried out under identical conditions to the first. This control is designed to show that any carbon dioxide produced not only comes from the peas, but is produced as a consequence of enzymatic reactions. In effect having the peas present, although dead, makes the control more nearly identical to the first experiment than would be the case if the peas were omitted.

(c) Bicarbonate indicator is sensitive to changes in pH. An intermediate red solution of the indicator becomes more yellow as the acidity increases and more blue as it decreases. If carbon dioxide is produced by the germinating peas it will make the solution more acidic because it combines with water to form carbonic acid which dissociates:

$$CO_2 + H_2O \rightleftharpoons H_2CO_3 \rightleftharpoons H^+ + HCO_3^-$$

carbon dioxide	water	carbonic acid	hydrogen ion	bicarbonate ion

In this case the bicarbonate indicator becomes yellow. If no carbon dioxide is produced, the indicator will remain unchanged and should any carbon dioxide be absorbed by the peas the indicator will become more blue. As carbon dioxide is produced by germinating peas the indicator in the experimental tube will become yellow. The dead peas should neither evolve nor absorb carbon dioxide and so the indicator in the control experiment should remain unchanged.

(d) Being holophytic (autotrophic) *Elodea* carries out photosynthesis, a process which uses some or all of the carbon dioxide produced during respiration. If carbon dioxide production during respiration by *Elodea* is to be measured accurately photosynthesis must be prevented. To do this the tube containing *Elodea* should either be placed in a dark room or be covered by lightproof paper. The wire gauze and handle can be dispensed with and the *Elodea* placed directly into the bicarbonate indicator solution, which is harmless to life. It is necessary to wash the apparatus and *Elodea* in bicarbonate indicator first as any trace of acid or alkali can cause a change in the colour of the indicator and mask other changes.

Example 8 (Time allowance 15 minutes)
Below is a table indicating the 'basal metabolic rate' (BMR) of a number of mammals:

Mammal	BMR (kJ m^{-2}) of body surface	Body mass (kg)
Shrew	1000	0·02
Rabbit	400	1
Sheep	180	50
Cat	320	3
Cow	90	400
Elephant	80	3600
Man	170	70

(a) Explain what is meant by 'basal metabolic rate'.
(b) (i) From the data, state the relationship between BMR and body mass. (ii) Explain this relationship.
(c) On the basis of your conclusions explain why
    (i) a reptile of body mass 1kg has a BMR of 15 kJ m^{-2} body surface.
    (ii) a bird of body mass 0·02kg has a BMR of 2000 kJ m^{-2} body surface.
(d) List **three** factors which can affect the BMR of an individual human.

Answer
(a) This is the minimum amount of energy required to maintain a homoiothermic animal in the proper condition, at complete rest.
(b) (i) In general the larger the body mass the smaller the basal metabolic rate. There are, however, deviations from this relationship due to other factors.
    (ii) The basal metabolic rate is measured when the organism is at complete rest. Energy is therefore only required to sustain the basic vital activities. As all animals in the table are mammals they are all endothermic (homoiothermic) and much of the energy used is to maintain a constant body temperature. A larger organism almost always has a smaller surface to volume ratio than a small one.

Heat loss from the body surface is consequently less compared to the smaller animal. The larger animal needs less energy to maintain its body temperature and its basal metabolic rate is correspondingly smaller.

(c) **(i)** This figure of 15 kJ m^{-2} body surface is lower than for a rabbit of the same mass which has a BMR of 400 kJm^{-2}. The reason is that whereas the rabbit is endothermic (homoiothermic) and needs energy to maintain its body temperature a reptile is ectothermic (poikilothermic) and has no such energy requirement. A reptile therefore has a lower BMR than an endothermic animal of corresponding mass.

(ii) The bird has the same body mass as the shrew and both are endothermic, and yet the bird has a BMR double that of the shrew. Birds have a high metabolic rate in order to allow the release of sufficient energy to permit flight.

Even at rest therefore, they have a higher metabolic rate than other endothermic animals of comparable size.

(d) 1 age 2 health 3 sex (male or female)

Example 9 (Time allowance 35 minutes)
Two experiments were conducted to measure the rate at which the spiracles of a locust opened and closed under various conditions of temperature and composition of inspired air. The following results were obtained:

Experiment 1		Experiment 2	
Temperature °C	Number of times spiracles opened per minute	Composition of inspired air	Number of times spiracles opened per minute
5	1	100% oxygen	2
10	3	Normal air	6
20	6	Air with 10% oxygen	9
30	8		
40	10	Air with 2% carbon dioxide	continually open

(a)   From the results obtained for experiment 1
(i)   calculate the temperature coefficient (Q_{10}) for the interval 10°–20°C.
(ii)   briefly describe the relationship between temperature and the activity of the spiracles.

(b) From the results obtained for experiment 2
(i) explain the changes in the rate of spiracle opening as the concentration of oxygen in inspired air changes.
(ii) how does raising the carbon dioxide concentration of inspired air to 2% affect the rate of spiracle opening?
(iii) suggest, with reasons, how the rate of opening and closing of spiracles is controlled.

Answer

(a) (i) The temperature coefficient is a measure of how many times an activity increases/decreases over a given temperature range. In this case it is how many times the opening of the locusts' spiracles increases or decreases between 10°C and 20°C.

At 10°C the spiracles open 3 times per minute
At 20°C the spiracles open 6 times per minute

The temperature coefficient (Q_{10}) therefore $= \dfrac{6}{3} = 2$

i.e. the rate of opening doubles over the temperature range 10°C–20°C.

(ii) The rate of spiracle opening increases with temperature although not uniformly. As the temperature increases from 5°C the rate of opening increases rapidly but above 20°C, although the rate of opening continues to increase, it does so less rapidly up to 40°C.

(b) (i) Although it is first necessary to establish the relationship between the concentration of oxygen in the air and the rate of opening of spiracles, the word 'explain' in the question indicates that reasons for the relationship are required. In 100% oxygen the spiracles open very infrequently (2 times/min.). With such a high concentration of oxygen (5 times normal atmospheric oxygen) the diffusion gradient across the spiracles is very large. Each time the spiracles open a large amount of oxygen enters which suffices for about 30 seconds before there is need to open the spiracles again to allow further oxygen to enter.

(ii) It causes the spiracles to remain open indicating that carbon dioxide concentration is the important factor controlling spiracle opening and closing.

(iii) Keeping strictly to the data there appear to be three factors that control spiracle opening and closing: the environmental temperature, the concentration of oxygen and the concentration of carbon dioxide.

A rise in environmental temperature causes an increase in

the rate of spiracle opening and closing. The reason for this is that in an ectothermic (poikilothermic) animal such as a locust, a rise in temperature causes a similar rise in all metabolic activity. The respiratory rate increases and consequently the need for more oxygen to be absorbed and more carbon dioxide to be removed. More frequent opening of spiracles allows this to take place. A rise in oxygen concentration decreases the spiracle opening rate whereas a fall increases it. This is because the higher the concentration gradient across the spiracles the more rapidly oxygen diffuses in. When oxygen concentration is higher than normal air the faster diffusion rate means the spiracles need open less often without reducing the total amount of oxygen diffusing in, in a given time. The reverse is true when the oxygen concentration is below that of normal air (e.g. 10% oxygen) and the spiracles need to open more frequently to supply the same amount of oxygen to the locust. Changes in carbon dioxide concentration have the most marked effect on spiracle opening and closing. A rise in carbon dioxide level to 2% causes the spiracles to remain permanently open. It must be remembered that this constitutes a rise of some 50 times from the normal carbon dioxide concentration in atmospheric air of 0·04%. Clearly such an increase in carbon dioxide concentration would normally only arise when carbon dioxide production by the locust was extraordinarily large, i.e. when the rate of respiration was exceptionally high. In these circumstances large quantities of carbon dioxide would need to be removed from the locust and equally large quantities of oxygen absorbed. The spiracles remain continually open to allow the oxygen and carbon dioxide to diffuse in and out respectively.

Example 10 (Time allowance 15 minutes)
(a)   Animals with lungs breathe the respiratory medium in and out through the same opening (= tidal flow) whereas fish, which have gills, pass their respiratory medium out through a separate opening. Suggest a reason for this.
(b)   Give **two** reasons why the absorption of oxygen by gills is more efficient in the osteichthyes (teleost fish) than in the chondrichthyes (elasmobranch fish).

Answer
(a)   Fish are aquatic organisms living in the relatively dense medium of water. Locomotion through such a medium is difficult. If water taken in during gaseous exchange were forced back out through the mouth (tidal flow) this water would tend to push the

68

fish backwards or at least slow down its forward movement. Such a slowing down might prove a disadvantage when the fish was chasing its prey or escaping predators. By pushing the water out behind it through the operculum or gill slits the flow actually assists the forward movement of the fish.

Air is far less dense than water and its expulsion has a negligible effect on locomotion. Terrestrial animals therefore expire the air through the same opening through which it was inspired. Being terrestrial the need to conserve water is important and restricting the openings of the body may have some small advantage in reducing water loss.

(b) In the osteichthyes the flow of water across the gills occurs in the opposite direction to the flow of blood through the gills (**countercurrent flow**). This method allows almost all the available oxygen in the water to diffuse into the blood. In the chondrichthyes however the water and blood flow across the gills in almost the same direction (**parallel flow**). This method allows only about 50% of the available oxygen to enter the blood. In the osteichthyes outward movement of the operculum during inspiration of water occurs. This means that for some part of the inspiratory period, pressure in the opercular cavity is below that in the buccal cavity and water flows over the gills. Water flows over the gills throughout the expiratory period. This means that water is flowing over the gills, and hence oxygen is being absorbed, for about 75% of the time. In the chondrichthyes there is no operculum or opercular cavity and water only flows over the gills during expiration. This means water flows over the gills, and hence oxygen is absorbed, for only 50% of the time.

Example 11 (Time allowance 25 minutes)
The lung volume of an adult human was measured by a spirometer during a variety of breathing exercises and the following trace obtained:

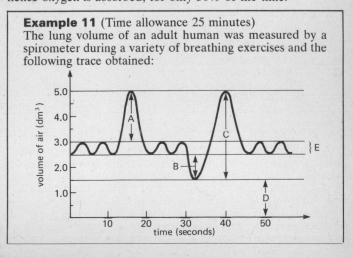

(a) By means of an appropriate letter state which volume represents
(i) inspiratory reserve volume. (ii) expiratory reserve volume. (iii) tidal volume. (iv) residual volume. (v) vital capacity.

(b) In another experiment a subject was given various gaseous mixtures to breathe and the rate of breathing was measured.
The results are shown graphically below:

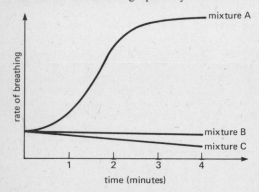

time (minutes)

Mixture A = 90% oxygen + 10% carbon dioxide
Mixture B = normal atmospheric air
Mixture C = 100% oxygen
(i) Explain what the data show concerning how the rate of breathing is controlled.
(ii) In the light of the information provided by the graph show why mouth-to-mouth resuscitation is a better means of artificial respiration than pressing on the chest wall.

(c) Why is it more dangerous to rebreathe expired air if it is passed through soda lime?

Answer

(a) (i) A (ii) B (iii) E (iv) D (v) C
(b) (i) Normal atmospheric air contains about 20% oxygen and 0.04% carbon dioxide. When the carbon dioxide concentration is increased to 10% (mixture A) the breathing rate increases and when it is decreased to 0.0% (mixture C) it decreases. This suggests that it is the concentration of carbon dioxide in inspired air that controls breathing. It is not the oxygen concentration in inspired air since when this is higher than normal at 90% it increases breathing rate and yet when

70

at 100% it decreases it. There does not therefore seem to be a logical relationship between the oxygen concentration in inspired air and breathing rate.

(ii) During mouth-to-mouth resuscitation expired air is forced into the patient's lungs. This expired air contains about 4% carbon dioxide and this stimulates an increase in the patient's respiratory rate and aids recovery. Pressing and releasing the chest wall causes atmospheric air with only 0·04% carbon dioxide to enter the patient's lungs. With its lower carbon dioxide level this air is not so effective in stimulating the patient's own respiratory rate and recovery is therefore slower.

(c) If expired air is rebreathed the oxygen content of it decreases and the carbon dioxide content progressively increases. The breathing rate is therefore increased due to the rise in the carbon dioxide concentration of the air. This faster breathing rate, to some extent, compensates for the lowering of the oxygen concentration and also acts as a warning to the person because the faster breathing rate causes distress. If the expired air is passed through soda lime, the carbon dioxide is completely absorbed and when rebreathed this air no longer stimulates faster breathing. The oxygen concentration of the air however continues to fall but there is neither a compensatory increased breathing rate nor a warning of the danger and so unconsciousness and death may follow.

Example 12 (Time allowance 5 minutes)
An athlete has a maximum air intake of 20 dm^3 min^{-1}. He can incur an oxygen debt of 14 dm^3 before he collapses. Running at 5 m sec^{-1} he uses oxygen at the rate of 0·3 dm^3 sec^{-1}. How far can he run before collapsing?

Answer
The air intake = 20 dm^3 min^{-1}, but only 20% ($\frac{1}{5}$th) of this is oxygen

Therefore oxygen intake = $\frac{20}{5}$ = 4 dm^3 min^{-1}

He can incur an oxygen debt of 14 dm^3
Therefore total oxygen capacity per minute = 4 + 14 = 18 dm^3
Using 0·3 dm^3 oxygen per second, in one minute he uses
$$0·3 \times 60 = 18 \text{ dm}^3$$
Therefore he can run for 1 minute before collapsing
At 5 m sec^{-1} he therefore runs
$$5 \times 60 = 300 \text{ m}$$

71

Chapter 8 Transport systems

Structure of blood

Example 1 (Time allowance 15 minutes)
The diagram below shows mammalian blood magnified about 1000 times.

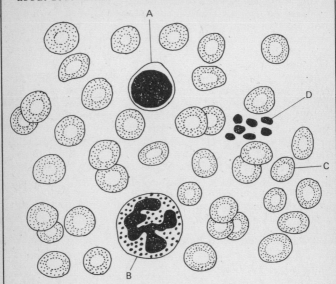

(a) Name the cells labelled A B C D.
(b) State the approximate numbers of each cell type in each mm³ of normal human blood.
(c) State one main function of each cell type.
(d) Outline the main stages in the process carried out by cell type D.

Answer
(a) A lymphocyte (agranulocyte)
  B polymorphonuclear leucocyte (granulocyte)
  C erythrocyte
  D platelet
(b) A 2500 B 7500 C 5 000 000 D 250 000
N.B. All figures are approximate since variations occur between individuals, e.g. between male and female.
(c) A production of antibodies
  B absorption of foreign material by phagocytosis

C   carriage of oxygen
D   clotting of the blood

(d) 1   cell type D + liquid from damaged cells $\longrightarrow$
                                                        thromboplastin
      2   thromboplastin + prothrombin (inactive) + vitamin K +
          Ca^{2+} $\longrightarrow$ thrombin (active)
      3   thrombin + fibrinogen (soluble) $\longrightarrow$ fibrin (insoluble)
      4   fibrin forms a meshwork which traps red blood cells thus
          preventing them escaping

Functions of blood

Example 2 (Time allowance 25 minutes)
The oxygen dissociation curves of human haemoglobin for a
normal person at rest at 37°C and for a human foetus, are
given below.

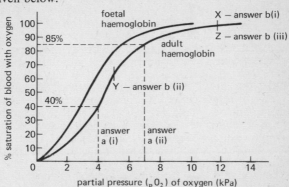

(a) State the % saturation of adult blood with oxygen
     when the pO_2 is:
     (i) 4 kPa   (ii) 7 kPa
(b) Using the letters X, Y and Z as indicated, mark on the
     graph for adult haemoglobin the positions of the
     oxygen tensions you would expect for the blood in:
     (i) the pulmonary vein (X)   (ii) the pulmonary
     artery (Y)   (iii) the femoral artery (Z)
(c) How would the dissociation curve differ if the blood
     had a high concentration of carbon dioxide?
(d) What is the significance of this difference?
(e) Explain why the oxygen dissociation curve for foetal
     haemoglobin is to the left of that for adult haemoglo-
     bin.

(f) Where would the dissociation curve for myoglobin be in relation to that for adult haemoglobin? Explain your answer.

(g) Where would the dissociation curve for the haemoglobin of a mud-dwelling organism such as *Arenicola* be in relation to that of an adult human? Explain your answer.

Answer

(a) (i) 40% (ii) 85% (Answers determined as shown on graph.)

(b) (i) The answer is marked X on the graph. The pulmonary vein carries oxygenated blood from the lungs where the pO_2 is about 13 kPa and the percentage saturation of haemoglobin is 95%. Since no oxygen is lost on its journey from the lungs to the pulmonary vein the oxygen tensions will be the same in both.

(ii) The answer is marked Y on the graph. The pulmonary artery carries blood that has returned from the tissues, via the heart, to the lungs. This blood is deoxygenated and has a pO_2 of about 5 kPa giving an oxygen saturation of around 66%.

(iii) The answer is marked Z on the graph. There should be no loss of oxygen between the lungs and the femoral artery as the blood does not pass through any tissue that absorbs O_2 from it during its journey. The oxygen tension should therefore be the same as that in the lungs (13 kPa).

(c) It would be of a similar shape but displaced to the right of the adult haemoglobin curve.

(d) The displacement to the right means that haemoglobin has a lower affinity for oxygen in the presence of high carbon dioxide concentrations. Since high carbon dioxide concentrations occur in respiring tissues haemoglobin readily gives up its oxygen at these tissues. The faster the tissue respires the faster the haemoglobin gives up its oxygen, which is precisely what is needed by a rapidly respiring tissue.

(e) Being to the left of the adult oxygen dissociation curve the haemoglobin of a foetus has a higher affinity for oxygen than the haemoglobin of its mother. This means oxygen will readily pass from maternal to foetal blood at the placenta.

(f) It would be displaced a long way to the left of the oxygen dissociation curve for adult haemoglobin which means it has a much higher affinity for oxygen. This is because myoglobin is found in muscle tissue where it can store oxygen for immediate use, and its higher affinity for oxygen allows it to readily absorb oxygen from the haemoglobin in the blood.

(g) It would be displaced to the left. In mud the organism lives in an environment of very low oxygen tension because oxygen is readily used by organisms breaking down organic material in the mud and there is little circulation of water to bring in more oxygen. If the organism did not have haemoglobin with the ability to absorb oxygen at very low oxygen tensions it would be unable to continue respiration aerobically.

Example 3 (Time allowance 5 minutes)
Study the equations below, all of which are involved in gaseous transport by the blood.
1   $H_2O + CO_2 \rightleftarrows H_2CO_3$
2   $H_2CO_3 \rightleftarrows H^+ + HCO_3^-$
3   $H^+ + HbO_2 \rightleftarrows HHb + O_2$
(a)   Which enzyme catalyses equation 1?
(b)   In which part of the blood do all three equations occur?
(c)   How does equation 1 affect the oxygen dissociation curve?

Answer
(a)   Carbonic anhydrase   **(b)**   In an erythrocyte.
(c)   The curve is displaced to the right and so haemoglobin has a reduced affinity for oxygen and so releases it more readily to the tissues.

Example 4 (Time allowance 1 minute)
Foreign proteins in the blood are called:
A   antibodies.   B   anti-toxins.
C   agglutinins.   D   antigens.

Answer   D
Anti-toxins (B) are produced by leucocytes to neutralize toxins produced by pathogens, agglutinins (C) are produced by the blood to cause coagulation of foreign material, and antibodies (A) are specific substances on cell membranes. B, C and A cannot be described as 'foreign'.

Blood groups

Example 5 (Time allowance 10 minutes)
When a sample of blood from each of five men was tested with blood typing sera anti A, anti B and anti D (rhesus) the following results were obtained as shown in the table overpage.

(a) Which man has blood containing the antigen A only?
(b) Which men are rhesus negative?
(c) Which man is blood group AB positive?
(d) Which man is termed the universal donor?

	anti A	anti B	anti D
Man 1	no change	agglutination	agglutination
Man 2	no change	no change	no change
Man 3	agglutination	no change	no change
Man 4	agglutination	agglutination	agglutination
Man 5	agglutination	agglutination	no change

Answer

(a) The blood should agglutinate with anti A serum only—man 3.

(b) Rhesus negative people show no agglutination when their blood is tested with anti-rhesus serum. These are men 2, 3 and 5.

(c) Man 4, who shows agglutination with all three sera.

(d) The universal donor possesses none of three antigens, A, B or rhesus, and so his blood may be given in small amounts to anyone. Such a person (group O) therefore has blood which fails to react (agglutinate) with any of the sera. This is man 2.

Blood circulatory systems

Example 6 (Time allowance 15 minutes)
The following diagram shows a plan of the mammalian circulatory system. The direction of flow of blood is indicated by arrows.

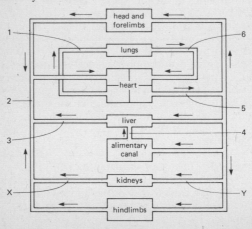

(a) Give the names of the six vessels indicated with numbers.
(b) State **four** differences between the composition of the blood in vessel X and its composition in vessel Y
(c) List **four** differences between the circulatory system shown and that of a fish.

Answer
(a)
1	pulmonary artery	4	hepatic portal vein
2	posterior (inferior) vena cava	5	aorta
3	hepatic vein	6	pulmonary vein

(b)   In vessel X there is more carbon dioxide, less oxygen, less glucose and more urea than in vessel Y. (There is also usually less water and fewer mineral salts in X, but these substances show much greater variation and it is therefore better to keep to the four main differences stated.)
(c)   (i)   There is a single circulatory system in a fish. (i.e. blood goes from the heart, through the gills and then to the rest of the body before returning to the heart.)
(ii)   The heart of a fish has only two chambers. (Not four as in a mammal.)
(iii)   There is a sinus venosus present before the auricle in fish.
(iv)   There is a renal portal system in fish. (i.e. blood from the posterior regions of the body returns via the kidneys rather than directly to the heart.)

Example 7 (Time allowance 10 minutes)
(a) State two differences between foetal and adult circulation.
(b) What is the importance of these differences to the foetus?

Answer
(a)   (i)   Foetal circulation has an opening (foramen ovale) between the left and right auricles of the heart. This is not in the adult circulation.
(ii)   There is a blood vessel (ductus arteriosus) joining the pulmonary artery and aorta in a foetus but not in an adult.
(b)   In a foetus the lungs do not function to obtain oxygen and there is no necessity to circulate more than a small amount of blood through them. The foramen ovale and ductus arteriosus divert most of the blood from the pulmonary to the systemic circulation, where it is most needed.

The mammalian heart

> **Example 8** (Time allowance 1 minute)
> The initiation of the mammalian heart beat occurs in:
> A   left auricle   B   aurio-ventricular node
> C   Purkinje fibres   D   sino-auricular node

Answer   D
The left auricle (A) plays no part in initiating the heart beat, the aurio-ventricular node (B) plays a direct role in the control of heart beat but does not initiate it and the Purkinje fibres (C) function to transmit impulses from the aurio-ventricular node to the apex of the ventricles.

Arteries, veins and capillaries

> **Example 9** (Time allowance 15 minutes)
> The table below contains data showing the capillary pressure in the lungs and body of a normal human. Study it carefully and answer the questions that follow.
>
> | | Lung capillary pressure (Pascals) | Body capillary pressure (Pascals) |
> |---|---|---|
> | Arteriole end of capillary | 5·3 | 30·5 |
> | Venule end of capillary | 2·1 | 10·2 |
>
> (a)   Explain how the heart produces a much lower blood pressure in the lung capillaries compared to those in the body.
> (b)   Give two reasons why lung capillary pressure must be low.
> (c)   How is water drawn back into the blood at the venous end of the body capillaries?

Answer
(a) The right ventricle, which pumps blood to the lung capillaries, has a thinner muscular wall than the left ventricle which pumps it to the body capillaries. The pressure created is directly proportional to the thickness of the wall of the ventricle from which it comes and hence the pressure is much lower in the lung capillaries.
(b)   (i)   The walls of the lung capillaries are extremely thin (0·2 μ m) and a low blood pressure is necessary if these walls are not to be damaged resulting in fluid loss.

78

(ii)   Gases need to be exchanged by diffusion between the air in the alveoli and the blood in the lung capillaries. This diffusion is more complete the longer the blood and air are in close contact. A low blood pressure means a slower movement of blood, a greater contact time between it and the alveolar air and consequently a better exchange of gases.

(c)   The osmotic potential of the blood is higher than the tissue fluid, due mainly to the blood proteins that were too large to enter the tissue fluid earlier. Water is therefore drawn back by osmosis. In addition there is a lower hydrostatic pressure in the blood due to the increased diameter of the venules compared to the capillaries. The high hydrostatic pressure of the tissue fluid therefore forces water back into the blood.

Example 10 (Time allowance 1 minute)
Small arteries have muscular walls because they:
A   give mechanical support to the thin wall of the artery.
B   help circulation by pumping blood.
C   allow the blood supply to an organ to be varied.
D   make the artery less permeable.

Answer   C
They vary the supply by constricting or dilating the artery according to the metabolic/respiratory needs of the tissue they supply. Mechanical support (A) is provided by collagen fibres not muscle. Pumping (B) is only carried out by the heart, though recoil of the artery wall may aid this. The recoil however is achieved by elastic fibres not muscle. Although the muscle may make small arteries less permeable (D) by increasing wall thickness, this is not *why* muscle is present.

Plasma, tissue fluid and lymph

Example 11 (Time allowance 1 minute)
Which of the following facts about the tissue fluid of mammals is true?
A   It has a higher osmotic potential than blood.
B   Only lymph vessels drain it away.
C   It is formed at the arterial end of a capillary.
D   It comprises more protein than blood.

Answer   C
It is formed at the arterial end of a capillary and the remaining options are incorrect because it has a *lower* osmotic potential than blood (A), the blood drains most tissue fluid away and only the residue is removed by lymph vessels (B), and most proteins are too large to enter the tissue fluid and so remain in the blood which consequently has *more* of them than does tissue fluid (D).

The uptake of water by plants

Example 12 (Time allowance 15 minutes)
(a) State two abiotic factors affecting water absorption by roots.
(b) State three functions other than absorption, performed by at least some roots.
(c) Describe two main pathways by which water passes from the soil to the root endodermis.
(d) What is the role of the Casparian bands in the passage of water through the endodermis?

Answer
(a) Temperature and osmotic potential of the soil solution.
(b) **(i)** Storage of food materials (e.g. in carrot roots)
(ii) Anchorage of plant (most, if not all, plants)
(iii) Reproduction (some roots are organs of vegetative propagation, e.g. *Dahlia*)
(c) **(i)** From vacuole to vacuole of adjacent cells, via the cell walls, membranes and cytoplasm.
(ii) Through the cell walls only of adjacent cells.
(d) They are impermeable to water and so all water is forced to pass through the cytoplasm of the endodermal cells rather than through the cell walls. Even in a plasmolysed cell the cytoplasm remains attached to the Casparian band and so water is forced into the cytoplasm. From the cytoplasm it is probably actively transported into the xylem.

Transpiration

Example 13 (Time allowance 20 minutes)
(a) Name the **two** principal conducting cells of the xylem.
(b) State **two** other cell types found in xylem and indicate their functions.
(c) What other two groups of substances are transported with the water in the xylem?
(d) Is transport in the xylem active or passive?
(e) What is the driving force that maintains transpiration and where is it located?
(f) The rate of transpiration usually exceeds the rate of water uptake during the day. Explain the importance of this in the mechanism of water transport in the xylem.
(g) Explain what is meant by root pressure and show how it can be demonstrated.

Answer

(a) (i) vessels (ii) tracheids

(b) **Name**   **Function**
    (i) xylem parenchyma   food storage
    (ii) xylem fibres   support

(c) (i) minerals
    (ii) organic substances (Sugars and amino acids are frequently, but not always, present.)

(d) Passive largely (Although movement from the endodermis to the xylem is active.)

(e) Evaporation of water from the leaves.

(f) It creates a pulling force drawing water up through the plant. As water evaporates from the leaves the cohesive properties of water draw up other water molecules. This pulling force means that water in the xylem is under tension provided the rate of transpiration exceeds the rate of water uptake.

(g) Root pressure is the force pushing water up through the plant. It is created by water entering root hairs by osmosis because the osmotic potential of the cells is greater than the osmotic potential of the soil solution. It is demonstrated by cutting a well-watered growing plant stem near its base. Water is seen to be continually exuded, showing that some force is pushing it up from the roots. It may be measured by cutting the plant under water and attaching a manometer to the cut end with a piece of tubing.

Example 14 (Time allowance 15 minutes)
Study carefully the data given below for four different species of plants.

Plant species	A	B	C	D
Relative number of stomata/mm² of leaf (upper:lower leaf surface)	5:30	0:80	10:15	0:50
Relative transpiration rate (upper:lower leaf surface)	10:12	0:4	15:30	20:50

(a) Comment on the distribution of stomata in each of the four species.

(b) Comment on the relationship between the distribution of the stomata and the transpiration rate in species B and D.

(c) From the data, what conclusions can be drawn about the differences in structure between the upper leaf surfaces of species B and D?

81

Answer

(a) **Species A** The stomata are fairly widely spaced on the lower surface compared to the other species but nevertheless six times more densely packed than on the upper surface.

Species B All stomata are on the lower surface where they are more densely packed than in any of the other species.

Species C The stomata are fairly evenly distributed between the upper and lower surfaces with half as many again on the lower surface. The stomata are more widely spaced on the lower surface than in any other species, but more densely packed on the upper surface than any other species.

Species D The stomata are confined to the lower surface only, where they are fairly densely packed, only species B showing a higher density.

(b) **Species B** The upper surface, in the absence of stomata, is completely impermeable to water. Surprisingly the lower surface, in spite of having the greatest stomatal density of any species, shows relatively little water loss through transpiration.

Species D The upper surface, although having no stomata, still shows quite a high transpiration rate. The lower surface has a rate much higher which can be accounted for by the relatively high density of stomata.

(c) Species B clearly has an upper surface impermeable to water whereas that of species D is highly permeable to it. The upper surface of species B probably has a thick waxy cuticle to prevent water loss whereas in species D it is very thin or absent altogether.

Example 15 (Time allowance 20 minutes)

Two different species of flowering plants growing in pots were subjected to still air conditions and the rates of transpiration for each was measured four times during the course of one hour. By means of an electric fan the same plants were then subjected to constant flow of air of 5 m s^{-1} and the transpiration rate was again measured four times during the next hour. The process was repeated in this way for wind speeds of 10, 15 and 20 m s^{-1}. The data obtained are set out below.

Wind velocity	Time (h)	Rate of transpiration (g^{-1}h^{-1})	
		Plant A	Plant B
Still air	0·25	200	60
	0·50	180	50
	0·75	200	60
	1·00	190	50

82

5 m s^{-1}	1·25	290	70
	1·50	300	80
	1·75	310	80
	2·00	290	70
10 m s^{-1}	2·25	400	100
	2·50	410	110
	2·75	400	110
	3·00	420	110
15 m s^{-1}	3·25	540	140
	3·50	530	140
	3·75	530	130
	4·00	460	140
20 m s^{-1}	4·25	420	170
	4·50	300	180
	4·75	210	180
	5·00	120	180

(a) Name the piece of apparatus most commonly used to measure water uptake in plants.

(b) Why is 'water uptake' measured by this apparatus not necessarily the same as the transpiration rate?

(c) Plot the graph of transpiration rate against time for both plant species on the same set of axes.

(d) What information is provided by the graph concerning the transpiration rate of the two species during the first three hours of the experiment?

(e) Explain the reduction in transpiration rate for species A after 3·75 hours.

Answer
(a) The potometer.
(b) Some water taken up will not be transpired, e.g. it may be retained in cell vacuoles to give turgidity, it may be used in growth or other metabolic activities such as photosynthesis or secreted by the plant (e.g. nectar). Some water transpired may not have come directly from water taken up during the experiment, e.g. it may be a metabolic product of reactions such as respiration, in which case it comes from stored food material made before the experiment began.
(c) In drawing the graph:
1 Choose the scale of each axis carefully to make maximum use of the graph paper.
2 Choose a scale that is easy to use and mark it clearly.

3 Name each axis clearly with the parameter being measured and the units used.
4 Plot the points initially in pencil so that errors may be easily erased. If necessary they may be later inked over.
5 Do not attempt to join up every adjacent point but draw a smooth curve that either joins points or passes close to them.
6 Distinguish the two lines drawn if possible (e.g. use a different colour for each) and in any case label them clearly with the letter of the appropriate species or give a key.
7 Mark the relevant point on the time axis where the wind speed is increased and state the velocity.
8 Give the whole graph a title.
It would be possible to draw a histogram but with so many points this would be more time consuming than a point graph.

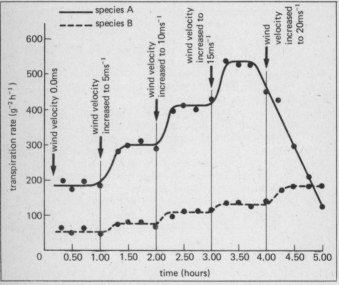

Transpiration rate against time for two species of flowering plants in different wind velocities

(d) The transpiration rate of both species increases initially as the wind speed increases but later stabilizes for a given wind speed with only small fluctuations. For both species the transpiration rate approximately doubles over the three-hour period. However, the initial rate for species A was almost four times that of species B. The difference in transpiration rate between the two species therefore doubles during the three hours.
(e) At this point the loss of water by transpiration for species A

84

far exceeds the supply of water to the plant. The plant wilts, i.e. the leaves lose turgidity and become flaccid. This reduces their surface area considerably with consequent reduction in the rate of transpiration. In addition loss of turgidity in the guard cells causes the stomata to close, considerably reducing the transpiration rate because most water loss occurs through open stomata.

The uptake of mineral salts

Example 16 (Time allowance 10 minutes)
The following graph shows the relationship between temperature and the uptake of potassium (K^+) ions by excised barley roots.

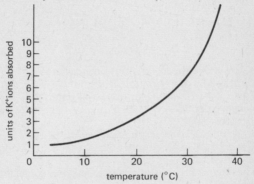

(a) Explain why the uptake of K^+ ions increases with a rise in temperature.

(b) What effect would the addition of potassium cyanide have on K^+ ion uptake? Explain your answer.

Answer
(a) A rise in temperature increases the kinetic energy of the K^+ ions which therefore diffuse in more rapidly, although not fast enough to explain the gradient of the graph. K^+ ions are also taken up actively, a process requiring energy. This energy is released during respiration and a rise in temperature increases the rate of respiration. With more energy available the rate of active uptake increases.

(b) Potassium cyanide would reduce K^+ ion uptake considerably. It inhibits the enzyme cytochrome oxidase at the end of the electron transport system of respiration. It therefore inhibits the whole of cellular respiration and prevents energy release. As K^+ ion uptake is partly active and requires energy the rate is reduced to that accounted for by diffusion only.

Translocation

Example 17 (Time allowance 1 minute)

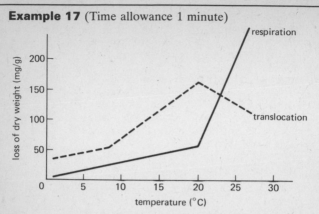

The graph above shows the relationship between temperature and the loss of dry matter from the leaves of a potato plant over a twelve-hour period in the dark. Which of the following best explains the fall in the rate of translocation at temperatures above 20°C?

A   Some energy from respiration is used in translocation.
B   The activity of sieve tubes decreases at temperatures above 20°C.
C   The movement of solutes is reduced after some hours in the dark.
D   Translocation and respiration use the same pool of metabolites.

Answer   D

The graph measures the loss of dry weight in the leaves. This loss will most probably be due to the removal of organic substances such as carbohydrates. Respiration, which uses carbohydrates, increases markedly at 20°C and so there is less carbohydrate to be translocated, i.e. the two processes use the same pool of metabolites. Option A may be discounted because if translocation used respiratory energy it could reasonably be expected to increase, not decrease, when the respiration rate rose. Sieve tube activity is unlikely to decrease above 20°C (B), especially when respiration is increasing above this temperature and a decrease in translocation after a period in the dark (C) does not explain the fall in translocation at temperatures above 20°C.

Chapter 9
Excretion and osmoregulation

Excretory products

Example 1 (Time allowance 15 minutes)
The periwinkle (*Littorina*) lives on rocky sea shores. Three different species each excrete their nitrogenous wastes in different forms as follows:
Species 1—ammonia
Species 2—urea
Species 3—uric acid
Suggest possible explanations for these differences.

Answer

Ammonia is a soluble but highly toxic excretory product which cannot be stored and must be continuously removed. Urea is also soluble but is less toxic than ammonia and therefore can be accumulated and stored before being excreted in a more concentrated solution. Uric acid is the least toxic of all and almost insoluble. It may be accumulated and removed in semi-solid form with the minimum amount of water. In effect, ammonia excretion involves much water loss, urea excretion rather less and uric acid excretion almost none at all. Organisms have evolved to excrete whichever substance is best suited to their needs in the environment in which they live. It must therefore be concluded that species 1 has little need to conserve water, species 2 has greater need to do so and species 3 must conserve it almost entirely. Because all the species exist on a rocky shore, there are no major temperature or other climatic variations between the habitat of one species and that of another. The over-riding factor affecting any shore is the tides. These follow an approximately twice-daily cycle during which the shore is covered and uncovered by sea water. Organisms high up the shore will be covered by water for much shorter periods than those lower down. When exposed the animals are subjected to desiccation. Those higher up the shore will lose most water and hence have the greatest need to conserve it, i.e. excrete uric acid (species 3). Those lower down have little need to conserve it and so excrete the simpler excretory product, ammonia (species 1). There is a monthly cycle of spring and neap tides. During neap tides the tidal range is small for a week or more when the species on the upper shore may never be covered by the sea. This reinforces the need to conserve water and so uric acid is excreted. Species low on the shore have a plentiful supply of water since they are only exposed for brief periods during low spring tides. It is therefore simpler for nitrogenous waste to diffuse out as ammonia than to carry out elaborate biochemical processes involved in changing it to one of the other excretory products. Species 1, therefore, probably occurs on the lower shore, or in rock pools, and species 3 occurs high up on the shore with species 2 living somewhere between the others.

Example 2 (Time allowance 1 minute)
Twenty-four hours after removing the liver of a mammal the concentrations of urea and amino acids in the blood would be changed in which of the following ways?
A   Both urea and amino acid levels would be greater.
B   Both urea and amino acid levels would be smaller.
C   The urea level rises while the amino acid level falls.
D   The amino acid level rises while the urea level falls.
E   There would be no change in the levels of either.

Answer   D
One major function of the liver is to deaminate amino acids and so form urea which can be excreted. With the liver removed the mammal would be unable to do this. Assuming all other organs function normally, amino acids will continue to be absorbed by the gut, but in the absence of the liver to deaminate them their level in the blood will rise. With no urea being produced by the liver but the kidney continuing to remove any urea present, its level in the blood will fall. Option D is hence correct.

Osmoregulation and excretion in animals

Example 3 (Time allowance 30 minutes)		
% concentration of sea water	**Number of vacuolar contractions/hour**	
	Species X	Species Y
0	100	25
5	90	22
10	80	65
15	70	62
20	60	53
30	35	28
40	20	13
50	6	4
60	0	0
70	0	0
80	0	0
100	0	0

Two species of *Amoeba* were placed in different concentrations of sea water and left for an hour during which the number of contractions of the vacuole were counted. The results are set out in the table.
(a) Plot these results graphically.
(b) Describe the relationship shown by the graph and give a full explanation.
(c) Describe how you would carry out the experiment.
(d) How could the results have been different if a small amount of mercury had been present in the sea water? Explain your answer.

Answer

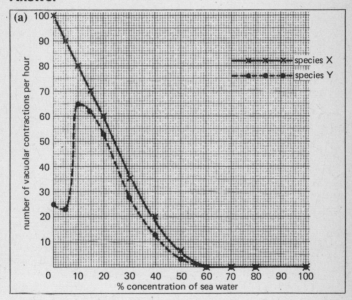

Graph of number of vacuolar concentrations against percentage concentration of sea water for 2 species of Amoeba

In drawing the graph attempt, as far as possible, to follow the basic guidelines below:
1 Plot the independent variable (or in this case percentage concentration of sea water) on the abscissa (horizontal axis) and the dependent variable (in this case number of vacuolar contractions per hour) on the ordinate (vertical axis).

89

2   Choose the scale of each axis so it makes maximum use of the graph paper and yet is easy to use.

3   Label each axis clearly with the parameter being measured and the units used.

4   Plot the points in pencil so that errors can be easily altered. If necessary the points can be made clearer later by inking over them.

5   Draw a smooth curve or line which is a reasonable interpretation of the points plotted. Do not necessarily join up each adjacent point, but draw the line reasonably close to each one.

6   Distinguish the two lines, if possible, by using a different colour or type of line (e.g. one continuous, the other dotted). Label each clearly with the letter of the appropriate species or use a key.

7   Give the whole graph a title. See the graph supplied.

(b)   For species X the number of vacuolar contractions decreases as the concentration of sea water increases until contractions cease altogether at 60% sea water. At this point species X has no need to remove excess water. It must be assumed that at concentrations of sea water less than 60% water enters species X by osmosis because its internal osmotic potential is greater than the osmotic potential of the surrounding sea water. This water needs to be removed by the contractile vacuole which periodically contracts emptying its contents to the outside. The more dilute the sea water, the greater the rate at which water enters and the more times the contractile vacuole contracts each hour. Species X shows no special ability to adapt to changing osmotic pressure in the medium around it, other than by removing excess water by alteration of the rate of vacuolar contraction. Species X has the same internal osmotic potential as a 60% solution of sea water (i.e. it is isotonic with 60% sea water). At sea water concentrations above 60% species X will progressively lose water because its internal osmotic potential is less than the surrounding medium. The contractile vacuole ceases to contract as this would only contribute to further dehydration. For species Y the number of vacuolar contractions falls slightly as the sea water concentration increases from 0 to 5%, but then increases rapidly for a further increase in sea water concentration. Above a sea water concentration of 10% the vacuolar contractions slowly decrease to 0 at a concentration of 60%. As with species X, species Y is isotonic with a 60% solution of sea water. Species X and Y therefore have very similar osmotic potentials. Unlike species X, however, species Y does not control water intake only by altering the rate of vacuolar contraction. At sea water concentrations of less than 10% species Y suddenly decreases its number of vacuolar contractions per hour. The organism is clearly not dead or dying because it starts to increase the rate of vacuolar contractions again at even lower sea water concentrations. As these contractions are an energy consuming process the organism must still be alive. Logically there are two explanations:

1   Water is entering more rapidly but the vacuole is failing to pump it out as fast as before.

2   Water is not entering as rapidly and therefore the necessity to remove it is less. The rate of vacuolar contractions falls.

If the first explanation were true the organism would eventually burst. In the absence of any evidence for this (e.g. the organism dying) it should be discounted. If the second explanation is true there are two possible reasons for water entering more slowly:

1   Species Y has lowered its internal concentration, possibly by losing some of its internal salts, sugars or wastes.

2   It has altered the permeability of its membrane to reduce water entry.

Whatever the reason, species Y is adapting to the new situation by reducing the necessity to remove water as rapidly. In this way it conserves energy. Species Y is more tolerant/better adapted to hypotonic environmental conditions than species X.

(c)   Large numbers of species X and Y should be obtained in separate containers. 1000cm^3 of sea water should be obtained and diluted as follows:

Concentration of sea water required	Amount (cm^3) of distilled water	Amount (cm^3) of normal (100%) sea water
0	100	0
5	95	5
10	90	10
15	85	15
20	80	20
30	70	30
40	60	40
50	50	50
60	40	60
70	30	70
80	20	80
100	0	100

About 5cm^3 of each dilution should be added to each of 12 watch glasses labelled with the appropriate sea water concentration. Too much sea water should be avoided otherwise the *Amoeba* become very diluted and difficult to find under the microscope at a later stage. Using a teat pipette a sample of species X should be added to each of the watch glasses taking care to transfer as little as possible of the liquid.

Transfer of too much of this liquid would alter the concentration of the solution in the watch glass. The samples should be left about 15 minutes to equilibrate and then a drop of each taken and transferred to a microscope slide and covered with a cover slip. The slide should be placed on a microscope stage and the microscope focused on one individual's contractile vacuole. Counting continuously for one hour would be laborious. It would be more practical to take 4 sample counts for 5 minutes, total the number of contractions and then multiply the result by 3 to give the number of contractions per hour. It would also be more representative if a different individual were used for each count. The procedure should be repeated for each of the dilutions and the results recorded. Finally the whole experiment should be repeated in the same way using species Y. All samples should be subjected to exactly the same environmental conditions (e.g. temperature, light intensity) and the same experimental techniques, otherwise the results will not be comparable.

(d) No vacuolar contractions would occur for either species at any dilution. Contraction of the contractile vacuole of all *Amoeba* species requires energy. This energy is released during respiration. Mercury (even in small quantities) is a metabolic poison inhibiting the enzymes involved in the hydrogen carrier (electron transfer) system of respiration. This system is therefore inhibited in the presence of mercury as in turn is the whole respiratory process. No energy is released and vacuolar contraction consequently ceases.

Example 4 (Time allowance 20 minutes)
An experiment was carried out to determine the effects of various treatments on the water loss from samples of twenty insects. Four desiccators were set up as follows:
A   insects with normal air.
B   insects with normal air and a dehydrating agent such as anhydrous calcium chloride.
C   insects dusted with an abrasive powder in normal air.
D   insects with air enriched in carbon dioxide.
During the next four hours the water loss was measured and the following results shown in the graph were obtained. Explain as far as possible these results.

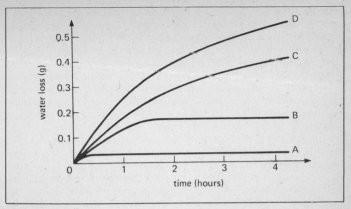

Answer

Insects are small terrestrial organisms whose large surface area to volume ratio means they easily become desiccated. To avoid this problem they have evolved a waxy waterproof cuticle which conserves water. All respiring cells need oxygen and hence the cuticle is not continuous over the whole body surface but is punctured at intervals by small holes called spiracles through which gases may diffuse. To minimize water loss these spiracles are surrounded by hairs and may be closed by means of valves when inward diffusion of oxygen is not necessary. With these facts in mind the water loss from the insect samples can be explained as follows:

Line A Insects with normal air The insects are in a desiccator which probably restricts their movement to some extent. Their respiratory rate will be relatively low as will their oxygen requirements. The insects start to lose a little water initially but then close their spiracles and reduce water loss considerably. As oxygen requirements are low the spiracles can be almost closed completely. Line A therefore flattens out. The small water loss that occurs could be through the cuticle and/or through the tiny aperture of the spiracles.

Line B Insects with normal air and a dehydrating agent such as calcium chloride The dehydrating agent dries the air surrounding the insects and the diffusion gradient of water across the spiracles is therefore greater. Water is lost more rapidly and line B rises more steeply. Water is also lost for a longer period of time. However, the need to conserve water eventually becomes paramount and so the spiracles are closed even more to prevent further desiccation. Line B flattens out but due to the drier atmosphere more water continues to be lost from the insects in group B than from those in group A.

Line C   Insects dusted with an abrasive powder, in normal air   As the insects move about their exoskeletons rub against the sides of the desiccator and each other. The abrasive powder scours the waxy cuticle and causes it to be worn away. Under these conditions the main means of preventing desiccation is removed and water is rapidly lost over the whole body surface. Line C rises rapidly and continues to do so up to four hours after the start of the experiment. The rate of increase in water loss, however, gradually slows down probably due to the insect having lost much of its water, thus raising its overall body osmotic potential and making further loss of water more difficult.

Line D   Insects with air enriched in carbon dioxide
The control of spiracle opening in insects is largely due to changes in carbon dioxide concentration although other factors such as temperature do affect it. A high respiratory rate causes an increase in carbon dioxide production by cells which builds up in the tracheae and causes the spiracles to open. Provided there is a higher than normal carbon dioxide concentration the spiracles remain permanently open. In air enriched with carbon dioxide this is clearly what has happened. The spiracles remaining open allow considerable water loss because the tracheoles which ultimately open to the spiracles are fluid-filled at their ends. The rate of water loss is very rapid at first, but as the insect becomes more dehydrated the osmotic potential of its tissues rises and so the rate falls because it becomes increasingly difficult to evaporate what remains.

Example 5 (Time allowance 15 minutes)
(a) Explain why adult frogs excrete urea and yet tadpoles excrete ammonia.
(b) Explain why the enzyme arginase in a frog only becomes active at metamorphosis.
(c) Explain why marine teleosts excrete small volumes of concentrated urine whereas freshwater teleosts excrete large quantities of dilute urine.

Answer
(a)   Ammonia is formed directly as a product of the deamination of amino acids. Ammonia is highly toxic to living cells and cannot therefore be accumulated, but must be removed rapidly. It is very soluble and so is excreted as a very dilute solution where sufficient water is available. Tadpoles live exclusively in freshwater and so water enters them by osmosis. It is a simple task to excrete the ammonia dissolved in this excess water which has to be continuously removed. Adult frogs live terrestrially for some of the time and so tend to dehydrate. The ammonia must hence be

94

removed by a method that conserves water. This method involves converting the ammonia to the less toxic urea according to the overall equation:

$$2NH_3 + CO_2 \rightarrow CO(NH_2)_2 + H_2O$$

ammonia   carbon dioxide   urea   water

The urea can be concentrated up to 2% before it becomes toxic and its removal therefore involves less loss of water for a given amount of nitrogenous waste removed.

(b) As explained in (a) the need for an adult frog to conserve water means that ammonia is converted to urea as the method of excreting nitrogenous waste, rather than excreting the ammonia directly as in a tadpole. Such a change in the excretory product produced takes place during the final stages of metamorphosis. The conversion of ammonia to urea takes place during the ornithine cycle. One enzyme essential to this cycle is arginase which converts arginine to urea. As urea formation only becomes necessary at metamorphosis, the arginase is inactive until then.

(c) The ancestors of present day marine teleosts arose in the sea but later invaded freshwater to which environment they became adapted. One such adaptation was to dilute their tissue fluids in order to reduce the tendency of water to enter by osmosis and flood them. Upon returning to the sea, however, the body fluids of these fish were hypotonic to (more dilute than) sea water and so they lost water by osmosis. They needed to conserve water and so they excreted urea rather than ammonia and concentrated it as much as possible, before eliminating it as concentrated urine.

The freshwater teleosts, however, have tissue fluids that are hypertonic to (more concentrated than) the water around them. Water enters continuously by osmosis. Urea was the excretory product used when they were marine and it has been retained despite changing to a freshwater environment. However, with excess water entering by osmosis their urea need not be concentrated and so is removed in the form of dilute urine.

Mammalian excretion

Example 6 (Time allowance 1 minute)
Which of the following would **not** normally make a human more thirsty?
A   drinking rum
B   drinking sea water
C   consuming large quantities of fat
D   sunbathing in hot weather
E   being diabetic

Answer C

The sensation of thirst occurs when the hypothalamus is stimulated by a rise in the osmotic potential of the blood. Any factor which therefore increases the osmotic potential of the blood will make a person feel more thirsty. Drinking rum (option A) increases the osmotic potential of the blood because it contains alcohol which is a diuretic, i.e. it causes an increase in the production of dilute urine. As more water is lost in the urine so the osmotic potential of the blood rises. Drinking sea water also raises the blood osmotic potential because the salt it contains must be removed to prevent upsetting the ionic balance of the body. Sea water is a 3·5% salt solution, but the body can only concentrate salt in the urine to levels of 2·2%. In other words removal of the salts involves more water loss than that provided by drinking the sea water. The blood osmotic potential therefore rises and thirst is experienced. Sunbathing in hot weather (option D) will induce sweating with consequent water loss, rise in blood osmotic potential and sensation of thirst. Being diabetic (option E) most often means that there is insufficient or no insulin produced in order to convert glucose to glycogen. The result is a rise in blood sugar level especially after consuming food. This causes a rise in blood osmotic potential and a feeling of thirst. Consuming large quantities of fat (option C) does not directly affect blood osmotic potential and so should not increase thirst, indeed its metabolism may even yield a little metabolic water, thus lowering the blood's osmotic potential.

Example 7 (Time allowance 20 minutes)

(a) List the ways in which a mammal obtains and loses water.

(b) A shipwrecked sailor in a boat on tropical seas has had nothing to eat or drink for four days. He drinks a small quantity of whisky he has with him but is still desperately thirsty. As a last resort he drinks large quantities of sea water. Next day he reaches land and quenches his thirst by drinking much freshwater. Describe the processes concerned with water balance that have taken place in the sailor during the events described.

96

Answer

(a) **Water obtained through**	**Water lost through**
Drinks	Urine
Food	Faeces
Metabolic by-product,	Sweating
e.g. respiratory product	Evaporation from lungs
	External secretions, e.g. tears

(b) Being in 'tropical seas' the sailor will be warm and sweating profusely (by day at least). With no major source of water to compensate for the loss by sweating the osmotic potential of his blood will rise. This rise will be detected by the hypothalamus as the blood passes through it, and the hypothalamus in turn will stimulate the pituitary to secrete antidiuretic hormone (ADH) into the blood stream. On reaching the kidney the ADH causes the loops of Henlé to reabsorb more water from the glomerular filtrate. This will reduce the amount of water lost in the urine and so help to conserve it. Drinking the whisky after four days only makes matters worse. The whisky contains ethyl alcohol which has three effects on the water balance of the body. Firstly, it inhibits ADH production from the pituitary and so allows more water to be lost in the urine than would otherwise be the case. In addition the alcohol is toxic and its removal (some is metabolized) involves its dilution in water, a further water loss. Thirdly, alcohol causes dilation of the blood vessels including the superficial ones of the skin. This may further increase sweating. In all, the whisky only increases the sailor's dehydration. Consequently the osmotic potential of the blood rises even more rapidly. The consumption of sea water is also detrimental. The salts taken in with the sea water must be removed if they are not to further concentrate the blood. Sea water is approximately a 3·5% solution of salts. At most the body can concentrate salts in the urine to 2·2%. This means the removal of the salt requires more water than the sea water provides and therefore dehydration becomes worse. The blood osmotic potential rises and more ADH is produced. Upon reaching the land the sailor quenches this thirst by drinking large quantities of freshwater. This is absorbed by the stomach into the blood and slowly the blood osmotic potential falls. The hypothalamus detects the change, the pituitary produces less ADH, the loops of Henlé gradually reabsorb less water and so the urine becomes increasingly dilute again, in order to maintain the normal blood osmotic potential.

Example 8 (Time allowance 30 minutes)
The following table shows the concentration of urine in different mammals relative to the concentration in man (the value for man is taken as an arbitrary unit of 1·00). The corresponding thickness of the medulla (inner region) of the kidney is given for each mammal. This thickness is measured relative to the size of the remainder of the kidney.

Mammal	Urine concentration (relative to man)	Thickness of medulla (relative to remainder of kidney)
Man	1·00	3·0
Otter	0·35	1·1
Sheep	0·70	2·2
Rat	1·92	5·4
Gerbil	4·15	9·1
Organism X	3·73	8·9

(a) Present these results in some suitable graphical form.
(b) State the relationship between the relative medulla thickness and the urine concentration.
(c) Give a possible reason for the relationship you described in (b).
(d) Suggest, with reasons, the possible habitat of organism X.

Answer

(a)

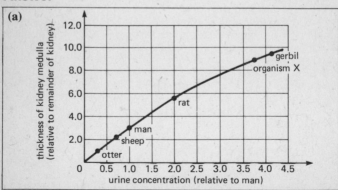

Medulla thickness against urine concentration for a variety of mammals

98

A cursory glance at the data should indicate that there is a direct relationship between the urine concentration and the thickness of the medulla. As one increases so does the other. A graph should therefore be plotted of one against the other, with each point marked with the name of the appropriate mammal. As both sets of data are variable it makes little difference which is plotted on the ordinate and which on the abscissa. A sketch of how the finished graph should look is shown.

(b) The relative medulla thickness and urine concentration show a direct relationship. In general, for any given increase in medulla thickness there is always a similar increase in urine concentration. In fact, the urine becomes increasingly concentrated for a given increase in medulla thickness and so the graph appears slightly curved, rather than a straight line.

(c) The vast majority of all water reabsorption from the renal fluid by the nephrons of kidney occurs as a consequence of the action of the loop of Henlé. This is a long hair-pin shaped structure, which, along with the collecting duct, forms a counter-current multiplier. This system allows progressive reabsorption of water as it passes down the collecting duct. The urine therefore becomes increasingly concentrated as it passes along the duct.

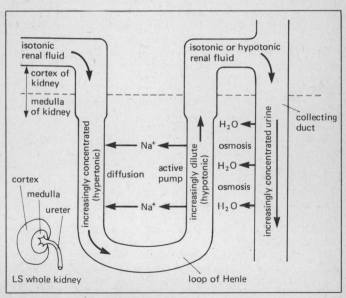

Diagram to show the counter current multiplier of the kidney

It follows from the diagram that the longer the loop of Henlé the greater is the distance along the collecting duct over which water is reabsorbed and hence the more concentrated is the urine produced. The loop of Henlé is mostly situated in the medulla (inner region) of the kidney. The longer the loop of Henlé, the greater the thickness of the medulla. In effect then the thickness of the kidney medulla is proportional to the length of the loop of Henlé which in turn is proportional to the urine concentration.
(d)   Organism X most probably lives in a habitat where water conservation is essential. For mammalian species, the majority of which are terrestrial, this is most likely to be a hot desert. The reasons why this is probable is that the medulla is relatively thick and the urine consequently highly concentrated, indicating conservation of water. The point for organism X on the graph lies close to that of the gerbil, which is a hot desert-living species that needs to conserve water. This lends further weight to the deduction that organism X has a hot desert as its habitat.

Example 9 (Time allowance 15 minutes)
(a)   The glomerular filtrate from the Bowman's capsule of the kidney is altered in composition as it passes through the proximal tubule. State three of these compositional changes.
(b)   If the proximal tubule were subjected to (i) cooling, (ii) cyanide, how would the composition of the urine produced be altered? Explain your answers.

Answer
(a)   1   The filtrate becomes less concentrated in salts.
      2   The filtrate becomes deficient in glucose.
      3   The filtrate loses a little water.
(b)   (i)   Cooling   The urine would probably consist of more salts than usual, and it would have some glucose present which is not normally the case. It would also be slightly more dilute. The reabsorption of glucose, salts and some water which takes place in the proximal tubule occurs against a concentration gradient and therefore involves active transport. The rate of active transport is directly reduced by cooling. In addition, cooling reduces the rate of respiration and as respiration provides the energy necessary for active transport its rate is further reduced. Diffusion still occurs but even the rate of this is reduced by cooling. It follows that less of the salt will be reabsorbed and its concentration in the urine will be greater than usual. Likewise the glucose may not be completely reabsorbed in the usual way and a little will be

100

present in the urine. Finally less water than usual will be reabsorbed and so the urine will contain more than normal. It is worth mentioning that water and salts are also reabsorbed further along the nephron; water from the collecting duct and salts from the distal tubule. It is possible, therefore, that provided they are functioning normally, these other sites of reabsorption will compensate for the reduced absorption in the proximal tubule, by absorbing more salts and water. The result would be urine with the normal concentration of these constituents.

(ii) Cyanide Cyanide inhibits the enzyme cytochrome oxidase which catalyses the final transfer of H^+ to oxygen during respiration. As such it acts as a respiratory inhibitor and prevents the release of energy. Because energy is essential for active transport, and absorption of materials by the proximal tubule is by this process, such absorption is reduced. It does not halt completely as cyanide does not affect diffusion which continues to account for some small amount of reabsorption. Again, therefore, the urine will contain some glucose, more salts and more water. If other sites absorb more water and salts to compensate for the slow absorption by the proximal tubule, the concentration of these in the urine may approach normal.

Chapter 10   Co-ordination

Neurones and the nerve impulse

Example 1 (Time allowance 1 minute)
The best description of a nerve impulse is:
A   the flow of electrons along the nerve axon.
B   a self-propagating polarity change along the membrane of a neurone.
C   the movement of sodium ions across the membrane of a neurone.
D   the sudden reversal of charges on the membrane of a neurone.

Answer   B

Multiple choice questions do not always have a single correct answer amongst three or four incorrect ones. Often all answers are correct in some degree and the candidate is expected to choose 'the most likely' or 'best' answer, as in this case. It is important that candidates consider all options carefully before deciding on an answer. Errors are likely if a candidate decides one answer is correct and so does not bother to read the alternatives or does so in a cursory manner. The best technique is to read all the options with an open mind and to reject them in order, starting with the least likely answer. Always reject answers on firm biological grounds and not because they do not sound right or you think it unlikely to be option A yet again. In the event of being totally unable to decide which of two alternatives is correct it always pays to guess unless the rubric clearly indicates that wrong answers are penalized. In this example option A is incorrect because nerve impulses are a consequence of ionic changes and not electron flow. Electron flow occurs in metallic conductors such as copper and is the basis of normal domestic electricity supply. Option C is an example of a partly correct answer which can be rejected because it is less complete than answer B. A nerve impulse does involve movement of sodium ions across the membrane, but in itself this movement does not constitute an impulse as there is not necessarily movement along the neurone. Indeed, sodium ions move across the membranes of all living cells, whether nerves or not. Option D is again partly correct but not as complete as B. Reversal of charges across the membrane of a neurone does occur during a nerve impulse but again the idea of this charge being self-propagating along the membrane is missing. This information is given only in option B which is therefore the 'best' description amongst the four alternatives.

Example 2 (Time allowance 1 minute)
In the following questions consider the four statements and decide which are correct and then give a single answer according to the following key:
Answer A if statements 1, 2 and 3 are correct
Answer B if statements 1 and 3 only are correct
Answer C if statements 3 and 4 only are correct
Answer D if statement 2 only is correct
Answer E if statement 4 only is correct
The speed at which an impulse is transmitted along a neurone is altered by:
1 whether the neurone is afferent (sensory) or efferent (motor).
2 the intensity of the generator potential.
3 whether the neurone is myelinated or not.
4 the axon diameter.

Answer C
Considering each statement in turn candidates should decide which are correct before going to the key to decide upon the appropriate letter. In this example statement 1 is incorrect because, providing all other factors are equal, impulses travel at the same speed along afferent and efferent neurones. Statement 2 is also incorrect as the speed of the transmission of the impulse is independent of the generator potential. Statement 3 is correct, impulses being transmitted faster along myelinated nerves because these neurones have points called nodes of Ranvier where the myelin sheath is missing. The action potential jumps from node to node which greatly increases the speed of transmission. Statement 4 is correct, the larger the diameter of the axon, the faster the rate of transmission of the impulse.
Summary: Statements 3 and 4 only are correct ∴ answer = **C**

Example 3 (Time allowance 10 minutes)
The table below shows the rate of conduction of nerve impulses along nerve fibres in three different animals.

Organism	Type of nerve fibre	Diameter of axon (μm)	Rate of conduction (m/s)
Squid	unmyelinated	1000	50·0
Frog	myelinated	10	20·0
Frog	myelinated	5	8·0
Human	unmyelinated	1	2·5
Human	myelinated	10	75·0

(a) What is the effect on the speed of impulse conduction of increasing the axon diameter?

(b) Suggest why myelinated fibres of 10μm diameter in humans should conduct impulses faster than 10μm diameter myelinated fibres of frogs.

Answer

(a) To give an accurate answer it is necessary to compare fibres of similar type, i.e. myelinated or unmyelinated and not compare one type with another. In this particular question it makes a difference to the answer obtained. The data provided clearly shows that the speed of impulse conduction increases as the fibre diameter increases in myelinated nerves. Likewise the impulse speed increases as the fibre diameter increases for unmyelinated nerves. However, a myelinated human nerve fibre of 10μm conducts faster than an unmyelinated squid nerve fibre which has a diameter a hundred times larger.

(b) As both organisms have myelinated nerves and they are of equal diameter, it is necessary to look at other differences between the two examples to explain the different speeds of impulse transmission. It should be obvious that any such differences must affect nervous conduction and the answer should explain how they have their effects. Possibly the most fundamental difference between a human and a frog is that humans are endothermic (homoiothermic) and frogs are ectothermic (poikilothermic). The body temperature of a frog rarely, if ever, exceeds that of a human. As a nerve impulse involves both diffusion and active transport of sodium and potassium ions, and as both processes are speeded by higher temperatures, it follows that the higher body temperature of humans will cause faster impulse conduction than in a frog, for a fibre of similar diameter and type.

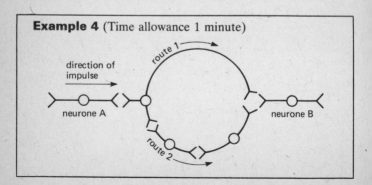

Example 4 (Time allowance 1 minute)

104

In the diagram shown, assume that the arrival of one
impulse at a synapse induces a single impulse in the
following neurone. Which of the following best describes
the impulses received by neurone B when a single impulse is
generated in neurone A?

A   two impulses, one stonger than the other
B   four impulses
C   one impulse
D   two impulses simultaneously
E   two impulses, one after the other

Answer E
Option A may be discounted because all impulses are of equal
magnitude, only the frequency of impulses varies. The impulse
from neurone A will divide and pass along routes 1 and 2 together.
There must therefore be two impulses reaching neurone B and
options B and C are incorrect. While transmission of an impulse
along a neurone is rapid, transmission across a synapse is
comparatively slow. As route 1 has no synapses apart from the
one preceding neurone B, and route 2 has two synapses over the
same distance, it follows that the impulse will take longer via route
2 than via route 1. The impulses will not arrive simultaneously
(option D) but one after the other (option E).

Example 5 (Time allowance 15 minutes)
Study the diagram below of a nerve synapse and answer the
questions that follow.

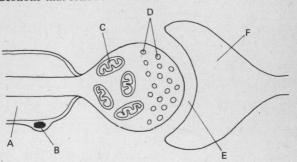

(a)   Identify structures A-F.
(b)   Why are structures C present in large numbers?
(c)   Describe the method of transmission of an impulse
       across the synapse.
(d)   State three possible functions of a synapse.

Answer

(a) A   myelin sheath   D   synaptic vesicle
    B   Schwann cell nucleus   E   synaptic cleft (gap)
    C   mitochondrion   F   axon of post-synaptic
                                             neurone

(b)   Mitochondria are responsible for releasing energy in cells.
During transmission of an impulse across a synaptic cleft the
chemicals in structures D are released and need to be resynthe-
sized before a new impulse can pass. The resynthesis requires
energy and hence structures C are present. Vital activities
frequently depend on a rapid succession of impulses and so
resynthesis needs to take place as soon as possible. Large numbers
of structure C ensure a greater release of energy and hence more
rapid resynthesis of the chemicals in D.

(c)   When an impulse travelling along the pre-synaptic neurone
arrives at the synaptic knob the synaptic vesicles (D) move
towards the pre-synaptic membrane. These vesicles contain a
transmitter substance (usually acetylcholine or noradrenalin)
which diffuses across the synaptic cleft (E) to the post-synaptic
membrane where special receptor sites receive the chemical. The
transmitter substance creates a change in permeability of the
post-synaptic membrane, thus allowing an influx of sodium ions.
This causes depolarization of the membrane which passes along
the post-synaptic neurone (i.e. a new impulse is initiated in the
post-synaptic neurone). The transmitter substance is broken
down by a specific enzyme, e.g. acetylcholine is broken down by
cholinesterase. The component parts of the transmitter substance
diffuse back through the pre-synaptic membrane and are
recombined using the energy provided by the mitochondria (C).

(d) 1   They act as valves in as much as impulses may travel
either way along a neurone but only in one direction across a
synapse.

2   They act as junctions. Impulses from many neurones may
meet at a synapse but only a single neurone may leave. Similarly a
single pre-synaptic neurone may create impulses along many
post-synaptic ones.

3   Synapses may 'filter' impulses. Impulses of low frequency
reaching a synapse may not initiate an impulse in the post-synaptic
neurone. Only when a certain threshold frequency of incoming
impulses is reached will a new post-synaptic impulse be initiated.
Low frequency impulses are thus 'filtered out'.

Example 6 (Time allowance 5 minutes)
A list of five regions is given through which the reflex arc of
a vertebrate passes, though not necessarily in the same
order as shown. Select one region from the list where each

---

of structures (i) to (v) is mainly found. Each region may be used once, more than once or not at all.

A spinal nerve
B ventral root of spinal nerve
C dorsal root of spinal nerve
D dorsal root ganglion
E grey matter of spinal cord
F white matter of spinal cord
    (i)   cell bodies of sensory neurones
    (ii)   axon of afferent (sensory) neurone
    (iii)   afferent (sensory) and efferent (motor) axons
    (iv)   axons that carry impulses to a muscle
    (v)   axons of intermediate (connector) neurones

Answer

(i)   D   These are grouped together in the dorsal root ganglion.

(ii)   C   The afferent neurones carry impulses towards the central nervous system along the dorsal root of the spinal nerve.

(iii)   A   Any nerve carrying both afferent and efferent neurones is called a mixed nerve. The only mixed nerve on the list is the spinal nerve.

(iv)   B   The impulses being carried to a muscle must originate in the central nervous system and the neurones whose axons carry them must therefore be efferent (motor) neurones. Such neurones are found mainly in the ventral root of the spinal nerve.

(v)   E   These occur in the grey matter of the spinal cord.

Sensory receptors

Example 7 (Time allowance 10 minutes)
One theory of colour vision suggests there are three different types of cone cell in the human retina, each containing a different variety of the colour-sensitive pigment, iodopsin. There are three varieties of iodopsin, one sensitive to red light, one to green and one to blue. The absorption of different wavelengths of light by the three types of cone is given on page 108.
(a)   From the data explain the following:
    (i)   Light of wavelength 430nm appears blue
    (ii)   Light of 550nm appears yellow
    (iii)   Light of wavelength 570nm appears orange
(b)   From your knowledge of the retina explain why two small objects close together can be more easily distinguished by cones than by rods.

Wavelength	Amount of light absorbed as a percentage of maximum		
(nm)	Red cones	Green cones	Blue cones
660	5	0	0
600	75	15	0
570	100	45	0
550	85	85	0
530	60	100	10
500	35	75	30
460	0	20	75
430	0	0	100
400	0	0	30

Answer

(a) **(i)** Looking at the table of data along the row opposite a wavelength of 430nm, it can be seen that no light is absorbed by either red or green cones. The blue cones, however, absorb their maximum amount of light. Only blue cones therefore initiate a nerve impulse under the influence of light of wavelength 430nm. On reaching the optic centre of the brain these impulses are interpreted as blue light.

(ii) The method here is the same, although the explanation is less simple. Looking along the row opposite 550nm it can be seen that red and green cones are stimulated equally (85% of their maximum absorption) but the blue cones are not stimulated at all. The brain therefore receives impulses of equal frequency from the red and green cones but no impulses from the blue ones. Red and green light when mixed in equal quantity produce yellow. The brain therefore interprets the impulses received as being yellow light.

(iii) Along the row opposite a wavelength of 570 nm it can be seen that the red cones have their maximum light absorption (100%), green cones have less than half their maximum light absorption (45%) and again the blue cones absorb no light (0%). The impulses received by the brain are therefore approximately ⅔ from red cones and ⅓ from green cones. The overall interpretation is orange.

(b) The cone cells of the retina each have their own independent neurone connected to the optic centre of the brain. Any two cone cells stimulated by the appropriate wavelength of light will each send impulses to the brain, which therefore interprets these as arising from two separate sources. Rods on the other hand are usually connected in groups to a single neurone. If any or all of the cells in one group are stimulated, only one neurone transmits impulses to the brain, which interprets this as arising from a single source. Light rays from two objects close together are likely to stimulate adjacent retinal cells. If these are cones, two separate images are perceived. If these are rods, only a single image is perceived unless the two rods belong to two different groups each with a separate neurone, although this is less likely to be the case. The situation is summarized in the diagram below.

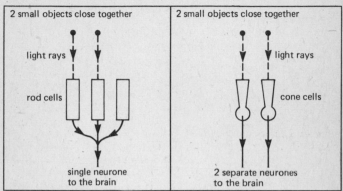

Example 8 (Time allowance 1 minute)
The refraction of light rays onto the retina of a vertebrate is mainly achieved by:
A   vitreous humour.
B   sclerotic.
C   iris diaphragm.
D   cornea.
E   suspensory ligaments.

Answer   D
The vitreous humour (A) is the fluid in the posterior chamber of the eye that helps to maintain the shape of the eyeball and so ensure that light rays are focused onto the retina rather than in front or behind it. It therefore aids focusing but is not the main means by which light is refracted. The sclerotic (B) is the tough fibrous layer that protects the eyeball. It is not transparent to light and therefore plays no role in refraction. The iris diaphragm (C) controls the amount of light entering the eye. It is again not transparent to light and plays no role in refraction. The suspensory ligaments (E) connect the ciliary muscles to the crystalline lens. They aid focusing by helping to change the focal length of the lens. They do not refract light and so can, like

options A, B and C be discounted. The remaining option, D, is the cornea, which is a transparent portion of the sclerotic at the front of the eye. It has a much higher refractive index than air and therefore is the main way by which light refraction is achieved.

Example 9 (Time allowed 4 minutes)
The tympanic membrane of the middle ear is 20 times greater in area than the oval window (fenestra ovalis). The two are joined by the three ear ossicles, two of which (the incus and the malleus) form an angled lever pivot. The distance from the tympanic membrane to the pivot is half as long again as the distance from the pivot to the oval window. Assuming that half the energy transmitted is lost due to friction, calculate the ratio of air pressure on the fenestra ovalis to that on the tympanic membrane.

Answer   15 : 1
If the tympanic membrane is 20 times greater in area than the fenestra ovalis, the pressure on the latter will be increased 20 times compared to that on the tympanic membrane. The pivot described has the longer arm towards the tympanic membrane and therefore any movement of this will be increased when transferred to the shorter arm, i.e. any pressure on the tympanic membrane will create an even greater pressure on the oval window. In this case the arm is 'half as long again' and so the pressure will be half as much again, i.e. it is increased from 20 times to 30 times. However, half the energy is lost as friction and so the air pressure on the oval window will be $30/2 = 15$ times greater than that on the tympanic membrane. The ratio of air pressure on the oval window to that on the tympanic membrane is therefore 15 : 1.

Temperature regulation

Example 10 (Time allowance 10 minutes)
(a)   Explain why a person with fever has a higher than normal body-core temperature and yet feels cold to the touch.
(b)   Why are there no adult mammals or birds that have a mass of less than 2 grams?
(c)   When the body temperature of a certain marine crustacean which lives under stones on the sea shore gets too high, why does it move out into the open where the environmental temperature is even higher?

Answer
(a) Fevers are ordinarily the result of infection. One effect is that the body temperature rises and this may mean that its immune responses, which are largely chemical, may act more efficiently. The body as a whole nevertheless reacts normally in trying to lower the temperature. One means of achieving this is by sweating. The water excreted from the sweat glands pours onto the skin surface from where it evaporates. In so doing heat energy is lost and this is taken from the skin which consequently feels cold, although the core temperature remains higher than normal.
(b) Birds and mammals are endothermic (homoiothermic) with constant body temperatures in the range 35–42°C. In almost all conditions they lose heat to the cooler environment. Very small organisms, even if basically spherical in shape, have a very large surface area to volume ratio. Heat is lost over the surface and so these small animals have problems in maintaining a constant body temperature. Even with insulation, a bird or mammal as small as 2 grams mass would be unable to maintain its body temperature and consequently its metabolic activity would be reduced until the organism eventually died.
(c) Crustaceans are ectothermic (poikilothermic) and their body temperatures are similar to those of the environment. To avoid denaturation of vital enzymes, however, they must control their temperatures within broad limits. It may seem an unusual reaction to move out from the shelter of a stone into a higher environmental temperature, in order to lose heat. It is usually very humid under stones on the sea shore and while crustaceans do not sweat, they can still evaporate moisture from their bodies. This evaporation is more rapid in the open, less humid atmosphere, than it is under the stone. The heat loss by this evaporation of water more than compensates for the higher environmental temperature and the crustacean loses heat overall.

Homeostasis and the endocrine system

Example 11 (Time allowance 25 minutes)
Two people drank a solution which contained 100g of glucose. The blood sugar level of each person was measured during the next three hours and the results are shown in the table on page 112.

(a) Using the same axes plot the blood sugar levels against time for X and Y.
(b) Summarize the information provided by the graphs.
(c) Suggest explanations for the changes in blood sugar levels of X and Y.

Time (mins)	Blood sugar level (mg/100cm³ blood)	
	X	Y
0 (glucose drunk)	81	90
20	136	131
40	181	142
60	213	89
90	204	79
120	147	74
150	129	86
180	113	89

Answer

(a) In plotting the graph it is essential to follow certain basic rules. These are laid out on pages 89 and 90. A sketch of how the finished graph should appear is laid out below.

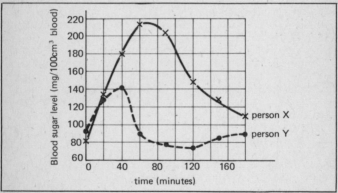

(b) Upon drinking a solution of glucose the blood sugar level of both X and Y rises immediately, person X showing the greater rise from 81mg/cm³ blood to peak of around 215mg/cm³ after 70 minutes. This is followed by a decrease to 113mg/cm³ blood after 3 hours. The **rate** of decrease slows over this period. Person Y appears to control blood sugar level to some extent. The rise is less pronounced (from 90 to 142mg/cm³ blood) and the fall in level begins earlier (after 40 minutes). The fall continues until the level is below the starting point after 2 hours suggesting that person Y has overcompensated for the rise in blood sugar level. The level rises, until after 3 hours it reaches the starting level again.

(c) The blood sugar level of person X rises almost immediately the glucose is swallowed. This is a result of absorption of the glucose into the blood by the wall of the stomach. As more and more glucose is absorbed, the blood sugar level rises steadily. After just over an hour most of the glucose has been absorbed. Throughout this period some of the glucose is used up in respiration and some has been excreted by the kidney when the blood sugar level became very high. These factors all combine to reduce the blood sugar level, rapidly at first, but more slowly as the level falls until the level approaches that before the glucose was swallowed. It seems unlikely from the data that person X produces much, if any, insulin and is therefore diabetic.

The blood sugar level of person Y rises rapidly after the glucose is swallowed again due to absorption by the stomach wall. After 40 minutes the level falls due to the rise in blood sugar level stimulating the secretion of the hormone **insulin** from the Islets of Langerhans in the pancreas. This hormone causes blood glucose to be converted to glycogen which is stored in the liver. Some use of glucose during respiration will also cause the blood sugar level to fall. As the level falls the secretion of insulin decreases. Due to the short time lag between the two events, by the time insulin production has ceased the blood sugar level has fallen below the original value. At this point the pancreas may produce **glucagon** which reconverts some of the glycogen back to glucose and the blood sugar level rises again. An equilibrium between the two hormones is established and the blood sugar level returns to the original value. Person Y shows a normal reaction in the circumstances.

Example 12 (Time allowance 1 minute)
Which of the following hormones will raise the level of lactic acid in the blood while decreasing the glycogen level of the liver?

A   adrenalin   D   insulin
B   cortisone   E   thyroxine
C   anti-diuretic hormone

Answer   A
Adrenalin (option A) has a number of effects on the body, all of which prepare it for some form of stress (e.g. a fight or flight situation). One such effect is to increase the rate of respiratory breakdown of glucose. This may occur before an increased supply of oxygen is available. In these circumstances, or if oxygen supply to the respiring tissues is otherwise inadequate, anaerobic respiration occurs and lactic acid is produced. This lactic acid diffuses into the blood causing its level there to rise. Another effect of adrenalin in preparing the body for stress and possible violent exercise, is the raising of the blood sugar level by converting liver glycogen to glucose. Adrenalin therefore both

113

increases lactic acid in the blood and reduces the liver glycogen level. Cortisone (option B) can be discounted because it **raises** the liver glycogen level. Anti-diuretic hormone (option C) helps regulation of blood osmotic potential and plays no part in regulating liver glycogen or blood lactic acid levels. Insulin (option D) raises liver glycogen levels and so is incorrect. Thyroxine (option E) controls metabolism and does not control liver glycogen or blood lactic acid levels.

Hormonal control of metamorphosis

Example 13 (Time allowance 15 minutes)
In a certain animal metamorphosis is controlled by two hormones A and B each produced by glands X and Y respectively. If gland X is removed from the animal a rapid succession of moults follows but these become only occasional on the reimplantation of gland X. The removal of glands X and Y together causes the complete cessation of moulting. From the information above:
(a) suggest the possible roles of hormones A and B in metamorphosis.
(b) suggest why some species of this animal continue to moult throughout their lives whereas other species cease to do so on reaching maturity.

Answer
(a) Gland X produces hormone A. When it is removed the absence of hormone A results in rapid moulting. It can be deduced that hormone A suppresses moulting. As gland Y, and hence hormone B, are still present, it must be assumed that hormone B is responsible for the rapid succession of moults. The reimplantation of gland X will result in the renewed production of hormone A. The effect is occasional moults, suggesting that hormone A only partially, not completely, suppresses hormone B. The removal of glands X and Y mean neither hormone is present. The absence of moulting confirms the view that it is a balance between hormones A and B that results in normal moulting frequency.
Summary: Hormone B causes a rapid succession of moults. Hormone A partially suppresses hormone B and results in occasional moults.
Hormones A and B act together to control moulting and hence metamorphosis.
(b) If we accept that hormone B causes rapid moulting and that hormone A suppresses it, it follows that any alteration in the balance between these two hormones will change the frequency of moulting. Those species that moult throughout their lives must

114

maintain production of hormones A and B at about the original balance. Those species that cease to moult at maturity must lose glands X and Y and hence hormones A and B, during the final metamorphosis and so cease to moult. This answer keeps strictly to the information given which is what the question **specifically** requires. However, it would be worth deducing from the information that metamorphosis could also be halted by increasing hormone A production, which suppresses hormone B, or by ceasing to produce hormone B which causes moulting.

Behaviour

Example 14 (Time allowance 10 minutes)
(a) List some differences between learned and instinctive behaviour.
(b) For the following examples state as precisely as possible the type of behaviour/response involved.
  (i) the closing of the Venus fly trap when an insect touches the hairs at the edge of the leaves
  (ii) the unicellular green alga *Chlamydomonas* moving to a warmer region in an aquarium

Answer

(a) Learned behaviour	Instinctive behaviour
1 Acquired during an animal's lifetime	Inborn, not acquired
2 May be intelligent. The animal often appreciates the function of the action.	Unintelligent with no appreciation of the function of the behaviour.
3 May easily and rapidly be adapted to suit changing circumstances.	Fixed pattern of behaviour. Only very minor modifications possible.
4 No fixed sequence of actions. The completion of one action need not necessarily affect the following action.	Often comprises a chain of actions, the completion of one triggering the start of the next.
5 Usually temporary form of behaviour, although it may be reinforced, making it more or less permanent.	The behaviour is permanent.
6 Varies considerably amongst different members of the same species.	Similar amongst all members of a species.

(b) **(i)** Thigmonasty **(ii)** Thermotaxis
'Thigmo' because touch is the stimulus, 'nasty' because only **part** of the plant responds and the direction of the response **is not related** to the direction of the stimulus.
'Thermo' because a temperature change is the stimulus.
'Taxis' because it is movement of the **whole organism** in response to a **directional stimulus**, the direction of the response being **related to** the direction of the stimulus.

Plant hormones

Example 15 (Time allowance 20 minutes)
When different concentrations of auxin were applied to the roots and shoots of a seedling of a certain species growth was either inhibited or increased. The effects are shown in the following graph.

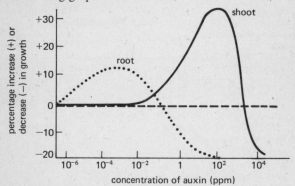

(a) What is the general relationship between auxin concentration and (i) root growth (ii) shoot growth?
(b) In view of these relationships suggest how auxins might normally be distributed in the seedling.
(c) Consider the following facts about auxins:
  (i) They are easily made synthetically.
  (ii) They are readily absorbed by the plant.
  (iii) They are not readily broken down.
  (iv) They are lethal to plants in low concentrations.
  (v) Narrow-leaved plants are more resistant to being killed by auxins than broad-leaved plants.
  Briefly explain the relevance of these facts to agricultural practice.

Answer

(a) **(i)** Low concentrations of auxin (10^{-6}–10^{-1}ppm) increase the growth of roots up to a maximum of 10%. Concentrations above 10^{-1}ppm inhibit growth to a maximum of -20% at 10^2ppm.

(ii) Auxin concentrations below 10^{-2}ppm have no effect on shoot growth. Above 10^{-2}ppm of auxin shoot growth increases up to a maximum of 30%. Above 10^3ppm of auxin growth is inhibited by up to 20%.

(b) It must firstly be appreciated that both the roots and the shoots of seedlings grow rapidly and mainly at the tips. As roots show the greatest increase in growth when auxin concentration is in the range 10^{-4} to 10^{-2}ppm it is reasonable to assume that this comparatively low concentration occurs at the root tip. Likewise in the shoot, maximum increase in growth occurs at auxin concentrations around 10^2ppm and this comparatively high concentration is most likely to be found at the shoot tip.

(c) The fact that auxins are readily absorbed, easily synthesized and are lethal to plants in low concentrations makes them useful herbicides. That they more readily kill broad-leaved plants than narrow-leaved ones is a particular advantage because many agricultural crops are narrow-leaved, e.g. cereals. These plants ordinarily have broad-leaved plants as weeds amongst them, which compete for light, water, nutrients, etc. Application of auxin at appropriate concentrations will kill only the weeds causing, at most, minimal damage to the crop. The fact that auxins are not readily broken down means that they persist in the soil and continue to act as selective weed killers for some time after application. This may mean that broad-leaved crops cannot be grown on that land for some time. There is, in addition, always the danger of accumulation in food chains where a herbicide is persistent. However, pure auxins, in the concentrations used, are not known to be toxic to animals.

Example 16 (Time allowance 1 minute)
When a young plant shoot is illuminated from one side, there is

A   growth only on the illuminated side.
B   growth only on the dark side.
C   more rapid cell division on the dark side.
D   more cell elongation on the dark side.

Answer   D
Plant shoots are positively phototropic and therefore bend towards the light when illuminated from one side. To achieve this, growth on the dark side must be more rapid than on the

illuminated side of the shoot. This allows option A to be discounted. The remaining options all produce bending towards the light. Option B suggests that there is no growth at all on the illuminated side. In practice the overall length of a shoot increases when showing a positive phototropic response. The bending is due to there being **more** growth on the dark side than on the light side. Bending always occurs just behind the tip in the region of cell elongation. Option C would produce bending at the tip because this is where cell division occurs. It can therefore be rejected. Option D satisfies both the need for growth on the dark side and in the region of cell elongation behind the tip.

Example 17 (Time allowance 15 minutes)
Three species of flowering plants were subjected to various periods of vernalization at 3°C. The time taken for each species to flower was recorded and the results obtained were plotted graphically.

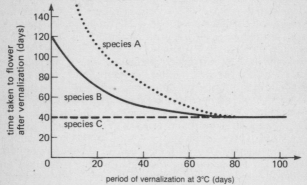

(a) Which species will only flower if vernalization has occurred?
(b) Which species does not require vernalization to accelerate flowering?
(c) Given non-flowering samples of each species what would be the minimum time of exposure to vernalization necessary to induce simultaneous flowering?
(d) One species is an annual, the other two species being perennials. Suggest, with reasons, which is the annual.

Answer
(a) **Species A** From the graph it can be seen that if the period of vernalization is less than 15 days, species A will not flower. (The line for species A becomes parallel to the flowering axis showing it

118

takes an infinite time to flower.) Species A must therefore have a period of vernalization of at least 15 days to induce flowering. The lines for the other two species cross the flowering axis, flowering occurs even when the period of vernalization is 0 days.

(b) Species C The line for species C is horizontal, crossing the flowering axis at 40 days. This shows that regardless of the length of the vernalization period species C takes 40 days to flower.

(c) 80 days This is the smallest value on the time of vernalization axis where all three lines for the three species virtually meet. For a period of vernalization of 80 days or more, flowering would occur in 40 days for all species.

(d) Vernalization is when a period of cold is necessary before flowering can occur. Perennial plants in temperate regions of the world naturally undergo such a period of cold during the winter months. The necessity of a period of vernalization thereby ensures that flowers are only produced after the winter, i.e. the flowers occur in spring and summer when insects are plentiful for pollination, rather than winter, when they are not. As an annual completes its whole life cycle in a single season, it follows that vernalization is not a requirement of annual plants so far as flowering is concerned. Species C is totally unaffected by changes in the period of vernalization (see answer to part (b)) and species C is therefore most probably the annual.

Example 18 (Time allowance 5 minutes)
State four differences between animal hormones and plant hormones.

Answer

Animal hormones	Plant hormones
1 The hormones are secreted by distinct specialized glands	There are no special sites for the synthesis of plant hormones, though synthesis is usually restricted to specific areas
2 Chemically the hormones are mostly steroids or polypeptides	Chemically the hormones are mostly indole- or purine-based
3 The hormones are transported by being pumped around the body in the blood	Transport of the hormone is largely by diffusion from cell to cell, though some transport in the phloem may occur
4 Transport of the hormones and hence their action, is relatively rapid	Transport of the hormones and hence their action, is relatively slow

Chapter 11
Movement, support and locomotion

Muscle structure and function

Example 1 (Time allowance 10 minutes)
(a) Give reasons why animals need to move from place to place.

Fig (a)

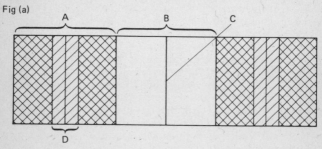

Fig (b)

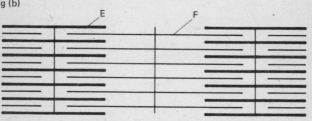

(b) The figure marked (a) shows part of a single myofibril from a striated muscle fibre. Below it, is the same portion shown in greater detail.
(i) Name the parts labelled A-F.
(ii) If the diagrams represent a myofibril in the relaxed state, how would they have differed if they had been drawn when contracted?

Answer
(a) It is always difficult in such a question to know exactly how many reasons are required or whether you have listed all the ones there are. Be guided initially by the time allowance when deciding how much detail is needed, but in any case keep the points general in a short answer question such as this. Only in an essay-type question would there be time to amplify the points in detail. Eight main reasons for animals moving from place to place are listed.

1 **To obtain food** The food requirements of most animals cannot be supplied by their immediate surroundings. Herbivores must move to new areas once the local supply is exhausted. Carnivores must hunt and capture their prey.

2 **To escape from predators** This is essential to survival.

3 **To find a mate** This is essential to the survival of a species. The ability to extend the range of potential mates ensures better mixing of genetic material (i.e. a larger gene pool), more variety of offspring and hence greater evolutionary potential.

4 **Better distribution** This prevents overcrowding, ensuring reduced competition for resources and a greater chance of survival.

5 **To find shelter** From both biotic and abiotic factors.

6 **Reduced vulnerability to disease** A scattered population is less likely to suffer epidemic diseases.

7 **Escape from waste products** These are toxic and may carry disease.

8 **To maintain position** A few animals such as sharks actually need to move horizontally to maintain a vertical position.

(b) **(i)** A   anisotropic band   D   H-zone (band)

              B   isotropic band   E   myosin filament

              C   Z-line (band)   F   actin filament

(ii) According to the theory of muscle contraction put forward by Huxley and Hanson the actin and myosin filaments slide further between each other during muscle contraction. If this is the case the H-zone and isotropic bands become shorter. The anisotropic band as a whole remains unchanged in size but the dark portion of it becomes larger. The sarcomere, i.e. the distance from one Z-line to the next, becomes shorter.

It would be possible to show these changes by means of a fully labelled and annotated diagram, although this could well take longer than an explanation in prose. The space on a structured paper left for the answer might give some clue as to the format required. A totally blank area would suggest a diagram was needed, lines indicate a written answer.

Bones, joints and muscles

Example 2 (Time allowance 15 minutes)

The figure represents a human leg with its bones and associated muscles. Each muscle is represented by a letter in a box and the points of attachment of the muscles are represented by arrows from the appropriate box.

(a)   (i)   What tissue attaches muscle to bone?

      (ii)   How is this tissue suited to its function?

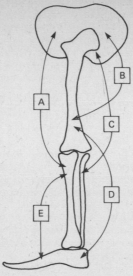

(b) Indicate by the appropriate letters which muscle or muscles are contracted when the following actions are carried out.
(i) The toes are lifted with the heel on the ground.
(ii) The knee is raised with the toes pointing downwards.
(iii) The heel is raised behind the knee, the foot hanging loose.

Answer

(a) (i) Tendon (white fibrous tissue).
(ii) Tendons comprise largely white fibrous tissue which is made of collagen fibres. These fibres are lined up longitudinally in the direction of most tension, i.e. along the line between the muscle and the bone. The collagen fibres are very strong and withstand tensions of up to 120 kg/cm^2. This suits them to the function of transmitting muscular force to the bone. This muscular force is sometimes considerable and so tendons must be able to withstand large strains. Collagen is inelastic. This means muscular movement will result in movement of the bone. Were it elastic the tendon would simply stretch when a muscle contracted and the bone would remain stationary.

(b) Before looking at the specific actions it is worth considering the following general points. Basically the muscles shown on the

122

diagram are attached between two bones. The contraction of any of these muscles will cause movement of the lower bone to which it is attached. All other bones below this lower point of attachment will also be moved. All muscles behind the leg (e.g. muscles B, C and D) when contracted will move the relevant part of the leg backwards. All muscles in front of the leg (e.g. muscles A and E) when contracted will move the relevant part of the leg forwards.

(i) To lift the toes with the heel on the ground involves the distance between the toes and the tibia (the front one of the two bones in the lower leg) becoming shorter. To achieve this would involve a muscle joining these two structures contracting. The muscle that is so positioned is E.

Answer   E

(ii) To point the toes downwards the heel must be pulled backwards and upwards. To achieve this, muscle D that joins the heel to the femur (bone of the upper leg) must contract while muscle E is relaxed. Raising the knee involves the distance between the knee and the pelvic girdle (hip girdle) shortening. The muscle that joins these two parts is muscle A. If this muscle were contracted the knee would rise. Altogether muscles D and A are involved with raising the knee with the toes pointing downwards.

Answer   D and A

(iii) If the foot is hanging loose it can be assumed that no muscle connected with the foot is contracted (namely muscles E and D). If the heel is raised behind the knee, the leg must be flexed at the knee. This involves shortening the distance between the lower leg and the posterior portion of the pelvic girdle. Contraction of muscle C would achieve this.

Answer   C only

Example 3 (Time allowance 25 minutes)
The diagram overpage represents a longitudinal section through a human femur.
(a)   Name the parts labelled A-E.
(b)   From the diagram show how the bone combines strength with comparatively little weight.
(c)   Explain how the structure of the bone drawn assists movement.

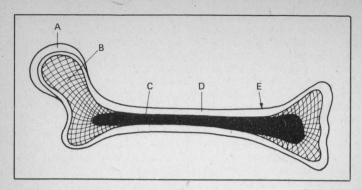

Answer

(a) A   articular cartilage
      B   spongy or cancellous bone
      C   marrow cavity
      D   compact (hard) bone
      E   periosteum (note the arrow head used to denote the
          outer surface layer)

(b) The bone is a human femur and as humans are bipedal it has to bear much of the body weight and is therefore usually under compression. In addition, during locomotion, the femur may also experience bending and torsion forces. These forces are most efficiently resisted by a cylindrical shape. Most of the stresses occur around the circumference of the bone, along the central region. By restricting hard bone, which is heavy, to these outer regions, the stresses are resisted adequately but weight is saved at the same time. At the ends of a bone forces act in a number of directions and the stresses are more widely distributed. No one place bears exceptional stress, but all parts bear some. The bone here need not be so compact, but it must be arranged in all directions. Spongy bone is used. It can withstand general stress in all directions but its honeycomb arrangement again reduces weight.

(c) It would be worth referring to part (b) first, and to explain how reduced weight combined with adequate strength assists movement. Clearly the skeleton must be carried around with the individual and so reduction in weight due to spongy bone and the marrow cavity is important. During movement, however, one femur may need to bear the whole of the body weight and so the cylinder of hard bone along the shaft is essential. The overall femur shape is important. The head is very rounded and forms the ball of a ball and socket joint. The round shape allows considerable freedom of movement. The limbs may be freely swung forwards and backwards to permit walking and running,

124

but at the same time allow some sideways movement to aid stability and other modes of motion such as climbing and swimming. The head itself is smooth to reduce friction and is protected by the articular cartilage (A) to cushion the impact of the head in the socket (acetabulum) of the pelvic girdle during walking or running. The other end of the femur has a hinge joint with the tibia of the lower leg. This gives flexibility in a forward and backward direction and aids most types of locomotion. An articular cartilage again cushions the bone.

The overall femur shape assists movement. The head is almost at right angles to the shaft. The femur stands out to the side of the body. This broadens the base upon which the body weight stands and gives much greater stability.

Locomotion

Example 4 (Time allowance 5 minutes)
At one stage during the locomotion of an earthworm the appearance of the body is as shown in the diagram below.

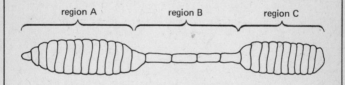

(a) (i) In which region or regions are the segments anchored to the ground?
     (ii) What structures are responsible for this anchoring?
(b) In region B are the circular muscles contracted or relaxed?
(c) What happens to the volume of a segment as it becomes elongated during locomotion?

Answer

(a) (i) At any one moment the segments that are anchored are those where the longitudinal muscles are contracted and the circular ones are relaxed (i.e. the 'short, fat segments'). Those correspond to **regions A and C.**
    (ii) **Chaetae** These are small bristle-like extensions that can be protruded or retracted from each segment, where they occur in paired groups on either side of the body.

(b) Circular muscles, as the name suggests, run in a circle around each segment. When contracted they get smaller in circumference, but being fluid-filled they are incompressible and so the volume remains the same. As the segment becomes smaller in circumference it must therefore become longer. This is the shape of the segments in region B so **the circular muscles must be contracted.**

(c) Each segment comprises various cells which are fluid-filled and a large cavity filled with coelomic fluid. These fluids are practically incompressible and therefore any constriction of one part produces a compensatory expansion in another part. Thus as the segment becomes smaller in circumference, it elongates. The volume of the segment remains the same.

Example 5 (Time allowance 10 minutes)

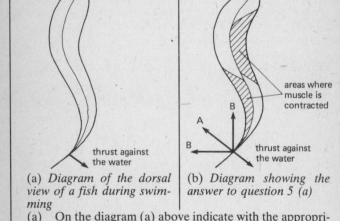

(a) *Diagram of the dorsal view of a fish during swimming*

(b) *Diagram showing the answer to question 5 (a)*

(a) On the diagram (a) above indicate with the appropriate labels the following:
  (i) the regions where muscle is contracted.
  (ii) the result of the reaction to the thrust against the water (indicate with an arrow labelled A).
  (iii) the resultant forces of this reaction to the thrust against the water (indicate with arrows labelled B).

(b) State which fins control rolling in fish and how they achieve this control.

(c) Why do the chondrichthyes (elasmobranchs) need to swim to prevent sinking whereas the osteichthyes (teleosts) can maintain position without swimming?

Answer

(a) The answer is shown in diagram (b).

(b) Rolling occurs when the fish rotates around its horizontal longitudinal axis. To control this rolling the appropriate fins should be placed to present the largest possible surface area in the direction of the rolling motion. In this position they present the greatest resistance to rolling because in order to rotate, large volumes of water would need to be displaced by the fins. In effect all the fins along the dorsal and ventral surfaces, and the paired pectoral and pelvic fins each help to control rolling.

Answer   dorsal, ventral, anal, pelvic, pectoral fins

(c) Osteichthyes possess an air-filled swim bladder which gives buoyancy. By adding gas to the bladder or by removing it the buoyancy can be regulated to suit different depths and conditions. In some fish this exchange of gas is achieved by swallowing air at the surface, in many there are special gas glands near the bladder which exchange gas between the swim bladder and the blood. Chondrichthyes have no swim bladder and their density is therefore greater than the water around them. Unless they swim and angle their fins in order to give lift, they sink.

Support in plants

Example 6 (Time allowance 15 minutes)

The figure represents the effect of wind on the stem of an herbaceous plant.

(a) Regions A and B are regions where certain strains develop if the wind causes bending of the stem. State the type of strain developed in (i) region A (ii) region B.

(b) The strains referred to in (a) occur largely at the periphery of the stem. Relate this fact to the internal distribution of tissues in a typical herbaceous stem.

(c) How does the arrangement of tissues in a typical herbaceous plant root differ from that in a stem? Relate these differences to the strains in a root.

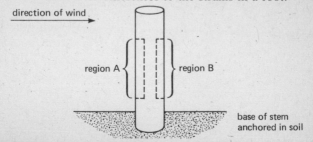

direction of wind

region A   region B

base of stem
anchored in soil

Answer

(a) **(i)** **Tension** The wind causes the stem to bend to the right. The left hand side of the stem as viewed in the figure therefore becomes longer. The cells in region A will therefore be stretched, i.e. under tension.

(ii) **Compression** As the stem bends the right hand side of the stem as viewed in the figure shortens. The cells in region B will therefore be under compression.

(b) While turgid parenchymatous tissue may give support, this is inadequate to resist the sudden and often strong bending forces created as a consequence of movement in the wind. These forces are resisted by the xylem of the vascular bundles, sclerenchyma and to a lesser extent collenchyma. The vascular bundles are arranged in a cylindrical pattern around the periphery of dicotyledonous herbaceous stems. While they are more scattered in monocotyledonous stems they are more frequent, and larger, towards the periphery. The sclerenchyma fibres are often found in association with the vascular bundles or in a band beneath the epidermis. In all cases this arrangement means that the tissues best able to resist tension and compression forces are located where these forces are greatest.

(c) In herbaceous plant roots the xylem that provides the main tissue resisting tension is located as a central core. Roots, being surrounded by soil, experience few lateral forces. The main forces are longitudinal, i.e. a pulling force. Such forces act centrally in the root rather than at the periphery. The central xylem column is therefore ideally placed to resist these forces.

Chapter 12 Growth and reproduction

Growth

Example 1 (Time allowance 10 minutes)
When a yeast colony was grown in a nutrient medium and the number of yeast cells counted over a period of 20 hours, graph 1 was obtained. If the rate of cell division was measured over the same period, graph 2 was obtained.

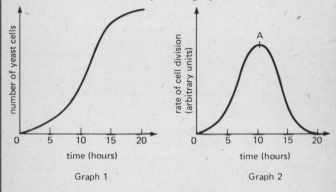

Graph 1   Graph 2

(a) Explain the shape of each graph.
(b) Suggest two possible reasons for the fall in growth rate after point A on graph 2.

Answer
(a) Graph 1 measures the actual number of yeast cells. As initial numbers are small, even when the cells are budding rapidly, growth is slow. However, as the number of budding individuals increases, numbers rise more rapidly and the graph increases exponentially. Some limiting factor causes the rate of increase to fall and so the graph begins to flatten until finally the number of new individuals exactly balances those dying, and so an equilibrium is established with no net increase in yeast cells. The graph flattens completely becoming parallel to the time axis.

In graph 2 this situation is more difficult and causes many students problems. It is essential to distinguish 'growth' and 'rate of growth'. If, for instance, the actual number of yeast cells is measured at five-hourly intervals and the results obtained are 100, 350, 900, 1600, 2150, 2400, 2400, when plotted on a graph, a shape similar to graph 1 is obtained = **growth curve**. If, however, the increase for each five-hourly interval (i.e. the difference between adjacent numbers) is calculated, the following is obtained.

Hours	0	5	10	15	20	25	30
Number of cells	100	350	900	1600	2150	2400	2400
Increase in cells at 5-hourly intervals		250	550	700	550	250	0

This last set of figures when plotted against time produces a graph similar in shape to graph 2 = **growth rate curve.**

Growth curve = amount of growth time^{-1}

Growth rate curve = amount of growth time^{-2}

The shape of graph 2 can now be explained. Initially when the number of yeast cells is small the rate of increase is low. As numbers increase so does the **rate** of increase until some limiting factor reduces the growth rate. Graph 2 therefore falls until the rate is equal to zero, i.e. actual numbers are not increasing. The growth rate curve is therefore a measure of the gradient of the growth curve.

(b) Choose from:

1 A build-up of waste products of the yeast cells, in this case ethanol, may begin to kill individuals.

2 The food supply (e.g. sucrose) of the yeast colony may be used up, so reproduction slows as less energy is available for budding.

3 An essential mineral such as nitrogen or phosphorus may be used up. Without these to produce the proteins for reproduction, the rate decreases.

In the absence of information regarding external changes such as a fall in temperature, it is better to restrict the choice of reasons to any two of the three above.

Mitosis, meiosis and reproduction

Example 2 (Time allowance 5 minutes)
The following diagram shows a cell with 8 chromosomes.

Indicate by means of further diagrams the appearance of the same cells during (i) metaphase of mitosis (ii) metaphase I of meiosis.

Answer

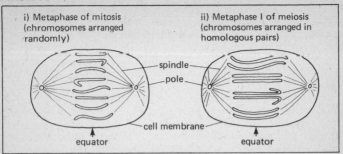

i) Metaphase of mitosis (chromosomes arranged randomly)

ii) Metaphase I of meiosis (chromosomes arranged in homologous pairs)

spindle

pole

cell membrane

equator

equator

In both diagrams it is important to show all 8 chromosomes and to draw them accurately. The spindle should also be shown and the nuclear membrane should have disappeared.

In the metaphase of mitosis the chromosomes should be lined up along the equator but not consistently associated in their matching pairs. In the metaphase of meiosis they should be lined up along the equator clearly associated in their matching (homologous) pairs.

Example 3 (Time allowance 10 minutes)
The following table gives the amounts of DNA in a cell at various stages of cell division. The least amount of DNA present at any stage is taken as 1·0 and this is used as a basis for comparison of the other stages.

DNA content of cell	Examples of stages of cell division
1·0	Meiosis, late telophase II
2·0	Mitosis, early interphase Mitosis, late telophase Meiosis, metaphase II
4·0	Mitosis, prophase Meiosis, anaphase I

Explain the differences in DNA content between:
  (i)   mitosis early interphase and mitosis prophase
 (ii) mitosis prophase and mitosis late telophase
(iii) meiosis anaphase I and meiosis metaphase II
(iv) meiosis metaphase II and meiosis late telophase II

Answer

(i) In mitosis early interphase, the chromatids disappear having completed a cell division. Towards the end of this division the chromosomes separated into their chromatids. During interphase these chromatids, which comprise the DNA of the cell, replicate. By the first stage of the next mitotic cell division (mitosis, prophase) the DNA content has therefore doubled from $2 \cdot 0$ to $4 \cdot 0$.

(ii) In mitosis prophase the chromosomes of a cell comprise two chromatids, one of each pair moving to an opposite pole. During late telophase the cell divides into two, thus halving the DNA content **per cell** compared to mitosis prophase, from $4 \cdot 0$ to $2 \cdot 0$.

(iii) Meiosis comprises a double division of the cell. Between the first meiotic division and the second meiotic division the chromosome, and hence DNA, content is halved. Anaphase I is followed by telophase I, where the actual division of the cell, and hence the halving occurs. No further division of the cell takes place before metaphase II and so the DNA content is simply halved, from $4 \cdot 0$ to $2 \cdot 0$.

(iv) Between these two stages the second meiotic division of the cell has occurred and so the DNA content has been further halved, in this case from $2 \cdot 0$ to $1 \cdot 0$.

Example 4 (Time allowance 20 minutes)
(a) (i) State three ways in which sexual reproduction may produce variation in offspring.
  (ii) What is the advantage of such variation?
(b) What is the main cause of variation during asexual reproduction?
(c) What are the main advantages of asexual reproduction?

Answer

(a) **(i)** The key word in this question is 'may'. The ways stated do not necessarily occur in all forms of sexual reproduction.
  1 Sexual reproduction involves the production of gametes which are often (though not always) produced by meiosis. During meiosis there is random distribution of homologous chromosomes on the metaphase plate. These chromosomes segregate independently of each other to produce a variety of gametes.
  2 During the extended first prophase of meiosis, crossing over occurs and genetic material is exchanged between homologous chromosomes. This increases variety.
  3 The gametes that fuse during sexual reproduction are often (though not always) from two different parents, each

132

with its own genotype. The offspring vary in that they bear some characters of each parent.

(ii) Evolution adapts a species to the prevailing conditions in order to ensure genetic survival. Conditions, however, change over a period of time due to climatic variations, geological changes, etc. It is in the interest of any species' survival to maintain a gene pool as varied as possible, in order that natural selection can choose those individuals best suited to any particular set of conditions. When conditions change, other individuals with different genotypes may be selected as being better suited to the new situation. The more varied the gene pool the more easily a suitably adapted organism can be selected. A second advantage of variation is that conditions not only change with time but also in space. Organisms may be dispersed to areas where different environmental conditions from those of the parents prevail. This is especially true for plants which, because they are dispersed by external agents, have little if any, choice in determining their destination. The more varied the offspring, the more likely it is that one or more individuals will be able to survive in the new environment.

(b) Mutations, both genetic and chromosomal.

(c) 1 It is rapid and can therefore be used as a means of building up numbers during favourable conditions.

2 Only one parent is required and no gametes are involved. The process is therefore simple, not involving complex mechanisms for bringing parents/gametes together. The offspring can be produced with greater certainty.

3 The offspring are often produced in large numbers and are easily dispersed (e.g. spores). They may also involve resistant/overwintering stages, e.g. perennating organs in plants.

4 The offspring are identical (mutations apart) to the parents. As most of these offspring will grow under the same environmental conditions as the parents, this means they have at least as good a genetic chance as the parents of surviving these conditions.

Sexual reproduction in flowering plants

Example 5 (Time allowance 10 minutes)
(a) List the differences between wind and insect pollinated flowers.
(b) Explain why it is incorrect to describe the flower of an angiosperm as its apparatus for sexual reproduction.

Answer
(a) Always be sure in this type of question to give *differences*. This involves stating clearly the contrasting features of both types

of flower. It is not sufficient to give a feature of one type, leaving the reader to infer the difference in the other type. The following table lists the major differences.

Wind	Insects
Pollen is small, light and smooth	Pollen is larger, heavier and may have projections
Pollen is produced in enormous amounts because there is considerable wastage	Less pollen is produced because pollination is more precise
Often in unisexual flowers (usually with an excess of male flowers)	Mostly in bisexual (hermaphrodite) flowers
Flowers dull, nectarless and scentless	Bright, scented flowers with nectar to attract insects
Stigmas long and protrude from the flower	Stigmas deep in the flower
Stigmas often feathery or adhesive	Stigmas not feathery, often small
Stamens long and protrude from the flower	Stamens deep in the flower

It is important to keep to features of 'flowers' as the question requires and not to include the plant as a whole.

(b) Plants show an alternation of generations between the gamete-producing gametophyte (sexual stage) and the spore-producing sporophyte (asexual stage). One adaptation of angiosperms to terrestrial life has been the evolution of a dominant sporophyte and almost the total elimination of the gametophyte, which becomes contained within the sporophyte. Because a flower is the reproductive apparatus of the sporophyte generation, it produces spores, not gametes, and is therefore asexual rather than sexual. These spores are of two types: **microspores** (pollen) and **megaspores** (one of which becomes the embryo sac). It is true that these spores contain nuclei which function as gametes, but the flower only produces these indirectly via the spores. It is therefore not wholly accurate to call the flower the apparatus for sexual reproduction, although it is easy to see how such a confusion can arise.

Example 6 (Time allowance 1 minute)
A pollen tube of a flowering plant is
A the male gamete.   D the female gametophyte.
B the embryo.   E a germinating spore.
C the male gametophyte.

Answer C
Option A may be discounted because the two male (sperm) nuclei of the pollen tube represent the gametes. The pollen tube is therefore more than just the gametes. Option B is incorrect because an embryo is the result of fertilization. However, the pollen tube contains the gametes **for** fertilization and must therefore precede fertilization rather than follow it. Option D is clearly incorrect. The pollen tube is a male structure not female. Option E could easily mislead candidates. The pollen tube does arise from germination of the pollen grain which is the microspore and might therefore be considered a germinating spore. The question, however, makes no mention of the spore (pollen grain) only the tube derived from it. This tube is the gametophyte because it is derived from a spore and it is male because it is derived from the male spore (microspore). Option C is therefore the best answer.

Example 7 (Time allowance 5 minutes)
Each of the lettered diagrams below are parts of a flowering plant, although not drawn to the same scale. The diploid number for this plant is 16.

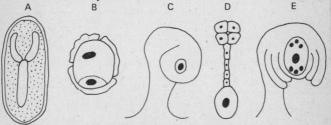

Give the letter which corresponds to the structure in which
   (i)   all the nuclei have 8 chromosomes.
   (ii) some nuclei have 8 chromosomes and the remainder have 16 chromosomes.
   (iii) some nuclei contain 24 chromosomes.
   (iv) a cell is about to divide meiotically.
   (v)   the male gametophyte is contained.

Answer
It is useful to identify the structures before attempting the questions. They are:
A   developing seed containing endosperm and embryo.
B   pollen grain.
C   early stage of ovule development.
D   the embryo: an octant, suspensor and basal cell.
E   mature ovule.

(i) The diploid number is 16. The structure in which all cells have 8 chromosomes must be haploid. While E has some haploid nuclei they are not **all** so. Structures A, C and D all have some diploid nuclei and no haploid ones. Only structure B, the pollen grain, has all haploid nuclei. **Answer B**

(ii) The structure required must have both haploid and diploid nuclei. This is structure E, the mature ovule. The three nuclei at each pole are haploid whereas the primary endosperm nucleus in the centre is derived from the fusion of two of the polar nuclei and is hence diploid. **Answer E**

(iii) At fertilization the diploid primary endosperm nucleus of the ovule fuses with one of the haploid sperm nuclei of the pollen tube. The resultant nucleus is therefore triploid (i.e. has 24 chromosomes). The material formed from such a fertilization is the endosperm and only structure A has this. **Answer A**

(iv) The cell about to divide meiotically is the megaspore mother cell shown in structure C. Any divisions about to occur in the cells of other structures will be mitotic. **Answer C**

(v) The male gametophyte is the pollen tube. Although not yet germinated it is contained in the pollen grain (structure B). **Answer B**

Example 8 (Time allowance 15 minutes)
The following chart provides information on seven plant species natural to Britain. It shows for each, when during the year leaves are present and when the flowers, if any, are produced.

Species	1	2	3	4	5	6	7
January							
February							
March							
April							
May							
June							
July							
August							
September							
October							
November							
December							

| leaves present | flowers present | fruits and seeds present |

For each of the following plant types state with reasons which of the species in the charts they are most likely to be:

(a) (i) a moss   (ii) deciduous trees (2 species)
     (iii) ivy   (iv) a fern   (v) a small woodland species
     (e.g. bluebell; lesser celandine)

(b) Of the four flowering species, which two are most likely to be insect pollinated? Give a reason for your answer.

Answer

(a) **(i)   Species 3**
Reason: mosses do not have flowers, fruits or seeds and it is therefore either species 1 or 3. As mosses do not lose all their leaves at one time but remain leafy throughout the year, this eliminates species 1.

(ii)   Species 2 and 4
Reason: deciduous trees lose their leaves at some time during the year. Species 3, 6 and 7 cannot therefore be deciduous trees. Almost all British deciduous trees bear flowers (larch is one exception). This eliminates species 1 which has none. Of the remaining ones, species 5 is the least likely to be a deciduous tree as it loses its leaves in mid June whereas most lose them during the autumn months. Species 2 and 4 are the most likely to be deciduous trees.

(iii)   Species 6
Reason: Ivy is an evergreen plant and hence must be species 3, 6 or 7. Ivy is an angiosperm and bears flowers. Of the three species mentioned only species 6 bears flowers.

(iv)   Species 1
Reason: Ferns are deciduous but do not have flowers. Only species 1 satisfies these conditions.

(iv)   Species 5
Reason: To obtain enough light to grow and produce flowers these plants emerge early in the year, before the canopy of the tree foliage has developed and shaded them. To achieve this early rapid growth they use carbohydrate stored in their bulbs (bluebells) or root tubers (lesser celandine) from the previous season's growth. As photosynthesis becomes more difficult due to lack of light when the taller plants and trees are in full foliage, they lose their leaves during mid-summer, soon after flowering. Species 5 is the only one that shows this type of life cycle.

137

(b) Species 2 and 5
Reason: Being ectothermic most insects do not emerge from over-wintering until the temperature is more favourable, i.e. March and April. Fewer insects are active in January and February (the main flowering period of species 4) or October and November (the main flowering period of species 6). Species 2 and 5 which flower in March and April are therefore more likely to be insect pollinated.

Sexual reproduction in animals

Example 9 (Time allowance 5 minutes)
Study the diagram below of a mammalian ovary.

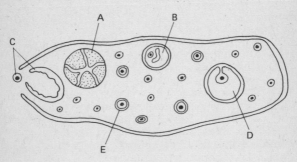

(a) Name the parts labelled A, B, C, D, E.
(b) State the correct developmental sequence of the structures labelled A, B, C, D and E.
(c) State whether each of the following is diploid or haploid.
(i) germinal epithelium (ii) ovum (iii) secondary oocyte (iv) primary oocyte

Answer
(a) A corpus luteum
    B primary oocyte surrounded by follicle cells
    C ovum being released from Graafian follicle at ovulation
    D mature Graafian follicle before ovulation
    E primary follicle

(b) The primary follicle (E) is present in the ovary at birth. After puberty it may mature into a primary oocyte (B) and later a Graafian follicle (D). The ovum from this Graafian follicle is released at ovulation (C) and the empty follicle develops into the corpus luteum (A).
The correct sequence is therefore E, B, D, C, A.

(c) **(i)** diploid **(ii)** haploid **(iii)** haploid **(iv)** diploid
The developmental sequence of the four structures above is germinal epithelium, primary oocyte, secondary oocyte, ovum. The first meiotic division which halves the number of chromosomes occurs in the formation of the secondary oocyte from the primary oocyte. The primary oocyte and the germinal epithelium from which it is derived are therefore diploid. The secondary oocyte and the ovum it develops into are haploid.

Example 10 (Time allowance 5 minutes)
The graph below shows the thickness of the uterus wall throughout the menstrual cycle in a human female.

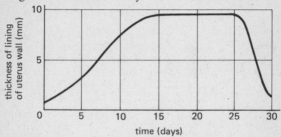

From the graph state the day
  (i)   ovulation is most likely to happen.
  (ii) assuming sperm are present, when fertilization is most likely.
  (iii) the corpus luteum begins to break down.
  (iv) menstruation begins.

Answer
N.B. The graph should be studied carefully before answering the questions. The answers should be obtained from the graph, *not from memory*. Most graphs use the start of the menstrual flow as day 1, whereas this graph uses the *end* of it as day 1. Memorized dates of the menstrual cycle may therefore give the wrong answers.

(i) In order to receive and nourish the newly-fertilized ovum, the ovary wall thickens and becomes better supplied with blood. It follows that ovulation should occur at a time that ensures that the ovum reaches the uterus when it is fully thickened. The journey from the ovary to the uterus may take a week but more commonly takes 5–6 days. Ovulation should therefore occur no earlier than 5 or 6 days before the uterus wall is fully thickened, although it could occur later. Ovulation could therefore occur between days 8 and 15 on the graph with **days 9 and 10** being the most likely.

(ii)   The ovum is usually fertilized 2 or 3 days after its release, which was established in part (i) as being most probably days 9 or 10. The most likely days for fertilization to occur are therefore **days 11, 12 or 13**.

(iii)   The corpus luteum breaks down shortly before menstrual flow begins. Since menstrual flow involves breakdown of the uterus wall, this can be seen on the graph to begin on day 25, because the thickness begins to decrease then. The corpus luteum must therefore begin to breakdown a few days earlier, e.g. **day 22 or 23**.

(iv)   This involves the breakdown, and hence decrease in thickness, of the uterus wall. According to the graph this begins on **day 25**.

Example 11 (Time allowance 20 minutes)
(a)   In a mammalian foetus there is an opening called the foramen ovale, between the left and right atria.
  (i)   What is the function of the foramen ovale?
  (ii)   It is normal for the foramen ovale to close at birth. If this did not happen what symptoms might be experienced by the baby?
(b)   State **three** features of the placenta that suit it to its function of exchanging materials.
(c)   Show how the placenta serves both as a link and as a barrier between mother and foetus.

Answer
(a)   (i)   As the lungs in a foetus do not function for gaseous exchange, forcing large volumes of blood over the pulmonary capillaries would be pointless and wasteful of energy. The function of the foramen ovale is to allow most of the blood to flow from the right atrium to the left atrium thereby by-passing the lungs and effectively making a single circulation in the foetus.
  (ii)   The foramen ovale is a 'hole in the heart' between the left and right atria. If it fails to close at birth the oxygenated blood on the left side of the heart is free to mix with the deoxygenated blood on the right side of the heart. The most important effect of this is that the blood in the systemic (body) circulation is only partially oxygenated. As partially oxygenated blood is bluer than if fully oxygenated, the overall appearance of the skin is bluer. The term 'blue baby' is often used to describe such a child. The presence of only partially oxygenated blood in the systemic circulation leads to inadequate oxygen supply to the tissues. This leads to

symptoms such as breathlessness and cramp especially during exertion.

(b) It is important to state features that suit the placenta to its function and not just state the functions. It is advisable to state briefly how it is related to its function.

The three best examples are:

1 It is composed of a large number of villi which provide a large surface area over which maternal and foetal blood come into close contact without mixing.

2 The distance between the foetal and maternal bloods is very small. As the rate of diffusion is inversely proportional to the square of the distance across which it takes place, this feature maintains a relatively rapid rate of diffusion.

3 The maternal and foetal bloods are kept moving across each other and this helps maintain a large diffusion gradient essential to rapid exchange.

(c) It is essential to the survival of the foetus that important materials and waste products pass between maternal blood and foetal blood, but at the same time materials that are potentially harmful to the foetus are prevented from crossing into the foetal circulation.

The placenta acts as a link because:

1 oxygen, water, soluble foods and salts pass from maternal to foetal blood.

2 carbon dioxide and nitrogenous waste pass from foetal blood to maternal blood for removal by the mother.

3 some of the mother's antibodies may pass into the foetal blood where they confer some immunity to the foetus.

The placenta acts as a barrier because:

1 it prevents maternal and foetal bloods mixing in case these are incompatible.

2 it prevents the higher maternal blood pressure being transmitted to the foetus.

3 it prevents the passage of some pathogens from the mother to the foetus.

4 it prevents the passage of those maternal hormones that could adversely affect foetal development.

Example 12 (Time allowance 1 minute)
For the following question, determine which of the responses that follow are correct answers to the question. Give the answer A, B, C, D or E according to the key below.
A   1, 2 and 3 are correct
B   1 and 3 only are correct

C   2 and 4 only are correct
D   4 alone is correct
E   1 and 4 only are correct
Direct effects of follicle stimulating hormone (FSH) include
1   development of the corpus luteum.
2   development of the Graafian follicle.
3   ovulation.
4   stimulation of sperm production.
5   stimulation of progesterone production.

Answer   C

The development of the corpus luteum (option 1) occurs
from the Graafian follicle immediately after ovulation. As
ovulation (option 3) is stimulated by luteinising hormone
(LH), it is LH and **not** follicle stimulating hormone (FSH)
that produces the corpus luteum—indeed it is from this very
effect that LH gets its name. Progesterone production
(option 5) is again stimulated by LH because progesterone
is produced by the corpus luteum. Development of the
Graafian follicle (option B) is stimulated by FSH as the
name suggests. What many candidates fail to appreciate is
that FSH is also found in males where it again stimulates
gamete (sperm) production. Options 2 and 4 are therefore
the only correct ones, making answer C correct.

Chapter 13   Genetics

> **Example 1** (Time allowance 30 minutes)
> A particular plant has two different varieties. If each variety is self fertilized, in both cases the offspring occur in the ratio 3 green to 1 white. On the other hand, cross fertilization of the two varieties produces F_1 offspring which are all green. Self fertilization of some of these F_1 offspring produces an F_2 generation in the ratio 9 green to 7 white but self fertilization of some other F_1 individuals gives all green plants. Assuming that the green colour is due to chlorophyll and its production is controlled in normal Mendelian fashion, explain fully these observations.

Answer

The fact that one of the F_2 generations produced offspring in the ratio 9 : 7 should suggest to candidates that this is a modification of the usual dihybrid ratio of 9 : 3 : 3 : 1. This means that the formation of chlorophyll in this particular plant is determined by two factors controlled by two different genes. Let these genes be A and B respectively. By normal Mendelian inheritance these genes each consist of two alleles, one dominant, the other recessive, i.e. A and B are respective dominant alleles, a and b the recessive ones. At chemical level each gene controls the production of a polypeptide which forms part or all of an enzyme. In this case it can be assumed that both polypeptides are required to form the enzyme(s) needed in the metabolic pathway leading to the formation of chlorophyll. Assuming that the polypeptide is only formed when the dominant allele is present, chlorophyll production requires alleles A and B. If either dominant allele is lacking chlorophyll is not formed and the plant is white.

The two different varieties when self fertilized produce three green offspring for each white one. The green offspring must have alleles A and B, whereas the white ones must lack one or both, i.e. **green plants must have the genotype A—B— and white plants must have the genotype aabb, aaB— or A—bb** (where — indicates either the dominant or the recessive allele). It is important to remember that the plants are **self fertilized**, i.e. the gametes are derived from the same genotype. To produce green plants this genotype must have the A allele and the B allele. In addition, to obtain white offspring it must have at least one recessive allele in order that a double recessive can occur. There are three possible genotypes that fulfill these criteria: AABb, AaBB and AaBb. The latter, however, when self fertilized, produces a 9 : 7 ratio of green to white (see later) and not a 3 : 1 ratio. This leaves the remaining two. As there are two different varieties that produce this 3 : 1 ratio when self fertilized, these must be the genotypes of these two varieties.

Proof

Variety 1 = AABb

Gametes = AB and Ab (Mendel's first law)
Self fertilization produces:

Gametes	AB	Ab	
AB	AABB	AABb	} F_1 generation
Ab	AABb	AAbb	

3 offspring have alleles A and B and are therefore green.
1 offspring (AAbb) lacks allele B and is therefore white.

Variety 2 = AaBB

Gametes = AB and aB (Mendel's first law)
Self fertilization produces:

Gametes	AB	aB	
AB	AABB	AaBB	} F_1 generation
ab	AaBB	aaBB	

3 offspring have alleles A and B and are therefore green.
1 offspring lacks allele A and is therefore white.
We are told that cross fertilizing the two varieties gives all green plants:
Variety 1 = AABb   gametes AB and Ab
Variety 2 = AaBB   gametes = AB and aB

Gametes	1 AB	1 Ab	
2 AB	AABB	AABb	} F_1 generation
2 aB	AaBB	AaBb	

All offspring have both allele A and allele B and are hence green.
We are told that self fertilization of some of these individuals produces plants in the ratio 9 green to 7 white, while self fertilization of others gives all green plants.
Examination of the four genotypes in the F_1 generation reveal two which were self fertilized earlier, i.e. AABb and AaBB.
In both cases self fertilization produced a 3 : 1 ratio of green to white plants. These two may therefore be discounted. One of the

144

remaining genotypes is AABB. This can only produce one type of gamete, namely AB. Self fertilization always results in a genotype AABB, i.e. all green plants. The remaining genotype is AaBb. This produces four different gametes, namely AB, Ab, aB and ab. Self fertilization produces:

Gametes	AB	Ab	aB	ab
AB	AABB*	AABb*	AaBB*	AaBb*
Ab	AABb*	AAbb	AaBb*	Aabb
aB	AaBB*	AaBb*	aaBB	aaBb
ab	AaBb*	Aabb	aaBb	aabb

All 9 offspring marked * have allele A and allele B which are needed for chlorophyll synthesis. These are therefore green. The remaining 7 offspring lack one or other dominant allele and cannot therefore synthesize chlorophyll. These are white.

Example 2 (Time allowance 15 minutes)
(a) A certain variety of mouse breeding true for long ears and black coat was crossed with another variety breeding true for short ears and white coat. The resulting F_1 generation all had an intermediate ear length and a grey coat. Explain this result.
(b) What would be the expected offspring in a cross between individuals of the F_1 generation?

Answer
(a) The presence in the F_1 generation of a character intermediate between the two parental characters always suggests incomplete dominance. In this case this would seem to apply to both the alleles for ear length and coat colour. In these situations one allele is equally dominant to the other at the same gene locus. It is normal to designate different letters to the alleles at this locus, rather than higher and lower case varieties of the same letter.

Therefore let allele for black coat = B
    ,,   ,,   ,,   ,, white coat = W
    ,,   ,,   ,,   ,, long ears = L
    ,,   ,,   ,,   ,, short ears = S

The mouse breeding true for long ears and black coat must be homozygous for both alleles, i.e. LLBB. Likewise the mouse breeding true for short ears and white coats must have the genotype SSWW.

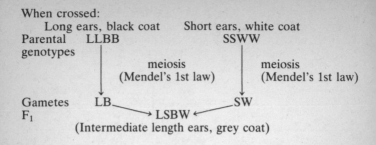

When crossed:

	Long ears, black coat	Short ears, white coat
Parental genotypes	LLBB	SSWW

meiosis (Mendel's 1st law)   meiosis (Mendel's 1st law)

Gametes   LB ⟶ LSBW ⟵ SW
F₁
(Intermediate length ears, grey coat)

(b) Individuals of the F₁ generation have the genotype LSBW (see part (a)).
The possible gametes from this genotype are LB, LW, SB, SW.

The possible offspring are therefore:

Gametes	LB	LW	SB	SW
LB	LLBB	LLBW	LSBB	LSBW
LW	LLBW	LLWW	LSBW	LSWW
SB	LSBB	LSBW	SSBB	SSBW
SW	LSBW	LSWW	SSBW	SSWW

The offspring are therefore:

Numbers	Phenotypes		Genotypes
	Ear length	Coat colour	
4	Intermediate	Grey	LSBW
2	Intermediate	Black	LSBB
2	Intermediate	White	LSWW
2	Long	Grey	LLBW
2	Short	Grey	SSBW
1	Long	Black	LLBB
1	Long	White	LLWW
1	Short	Black	SSBB
1	Short	White	SSWW

146

> **Example 3** (Time allowance 10 minutes)
> In birds barred plumage is a sex-linked dominant character. A cross between a cock without barred plumage and a hen with barred plumage produced an F_1 generation of barred cocks and non-barred hens and an F_2 generation of barred and non-barred individuals of both sexes, in equal numbers. Explain this.

Answer

It is most important to remember that in birds the male is the heterogametic sex (i.e. XY) and the female is the homogametic sex (i.e. XX).

Let the allele for barred plumage $= B$

,, ,, ,, ,, non-barred plumage $= b$

As almost all sex-linked alleles are carried on the X chromosome the alleles are represented by X^B and X^b respectively. Non-barred cock therefore $= X^b X^b$ (no other genotype is possible as we are told that barred plumage is dominant).

Likewise barred hen $= X^B Y$.

Parents $X^b X^b$ $X^B Y$

                          meiosis

                (Mendel's 1st law)

Gametes X^b X^B and Y

F_1 generation:

♀↓ Gametes ♂→	X^b	X^b
X^B	$X^B X^b$	$X^B X^b$
Y	$X^b Y$	$X^b Y$

The offspring are therefore equal numbers of barred cocks ($X^B X^b$) and non-barred hens ($X^b Y$).

To obtain the F_2 offspring:

Parents $X^B X^b$ $X^b Y$

                        meiosis

               (Mendel's 1st law)

Gametes X^B and X^b X^b and Y

♀↓ Gametes ♂→	X^B	X^b
X^b	$X^B X^b$	$X^b X^b$
Y	$X^B Y$	$X^b Y$

147

Numbers	Phenotype	Genotype
1	Barred cocks	X^BX^b
1	Non-barred cocks	X^bX^b
1	Barred hens	X^BY
1	Non-barred hens	X^bY

These ratios are those given in the question.

Example 4 (Time allowance 20 minutes)
In *Drosophila* a group of flies known to be homozygous recessive for two genes A and B were crossed with a group known to be heterozygous for both genes. The resultant adult offspring are as follows:
Genotypes: AaBb   aabb   Aabb   aaBb
Numbers: 306   295   107   102
Explain these results as fully as possible.

Answer
The cross between individuals homozygous recessive for both genes (i.e. aabb) and ones heterozygous for both genes (i.e. AaBb) should produce offspring of the genotypes shown above, but in approximately equal numbers:

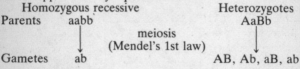

Homozygous recessive   Heterozygotes
Parents   aabb   AaBb
                        meiosis
                 (Mendel's 1st law)
Gametes   ab   AB, Ab, aB, ab

F_1 generation:

Gametes from homozygous recessive individual ab	Gametes from heterozygous individual			
	AB	Ab	aB	ab
	AaBb	Aabb	aaBb	aabb

The predicted ratio is therefore 1 : 1 : 1 : 1.
In practice, however, it is:

Genotypes	AaBb	aabb	Aabb	aaBb
Predicted ratio	1	1	1	1
Actual numbers	306	295	107	102
Approximate ratios of actual numbers	3	3	1	1

The most likely explanation for the actual ratio is that genes A and B occur on the same chromosome, i.e. are linked. If this were the case, then assuming AB and ab occur together on respective homologous chromosomes, they would tend to remain together, i.e. the heterozygote (AaBb) would only produce two types of gamete, namely AB and ab.

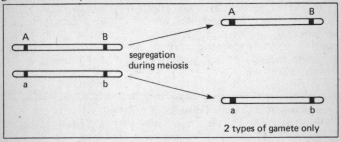

Because A and B/a and b are linked, they do not segregate independently. When each of these gametes fuses with the ab gamete from the homozygous individual, two types of offspring result—AaBb and aabb. While these represent the majority of the offspring produced in the answer given, they are not the only types produced. Genotypes Aabb and aaBb must therefore be a result of crossing over during meiosis.

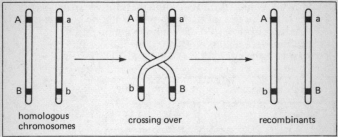

As a result of such crossing over, two other gametes arise, namely Ab and aB. When fused with the ab gamete from the homozygous recessive parent the two other genotypes Aabb and aaBb are produced.

If cross over occurred every time the gametes were formed, only gametes Ab and aB would occur. If it occurred 50% of the time gametes Ab, aB, AB and ab would occur in equal numbers. As it is gametes AB and ab occur three times as frequently as gametes Ab, aB, i.e. they represent 75% of the total number of genes and the recombinants represent 25%. Cross over must occur 25% of the time, i.e. once for each four gametes produced. Other possible explanations are given over page.

149

1   It is only a statistical difference and the four types of gamete are produced in equal numbers, but by chance gametes ab and AB happen to fertilize more often than gametes Ab and aB. Given the size of the offspring samples, statistical error seems unlikely.

2   The four types of gamete are produced in equal numbers and fuse with the ab gamete in equal numbers, but offspring of genotype Aabb and aaBb are selected against during development. By the time the flies are adult these genotypes are considerably under-represented in the sample. Although a possible explanation, the absence of any information about selection means it is best only to make a brief mention of this possibility.

Example 5 (Time allowance 5 minutes)
In a paternity suit, a mother of blood group A has a child of group O. The man she claims to be the father is group B. Explain whether this information can settle the issue one way or the other.

Answer
In humans, blood groups are controlled by three alleles A, B and O occurring at the same locus of a chromosome. Alleles A and B are equally dominant, and both are dominant to allele O. The mother who is blood group A, could therefore be of genotype AA or AO. Likewise the father (group B) could be genotype BB or BO. The child (group O) can only be genotype OO.

If the father were genotype BO and the mother AO, then a child of group O (genotype OO) could result.

Father's gametes →		B	O
Mother's	A	AB	AO
gametes	O	BO	OO

Indeed all possible blood groups could arise from this cross. The issue cannot be settled because, while the man could in theory have fathered the child, so could any fertile male carrying an O allele. This includes everyone of genotype AO, BO or OO. In Britain more than half the population have one or other of these genotypes.

Chapter 14   Evolution

Example 1 (Time allowance 1 minute)
When formulating his theory of evolution Darwin consi-
dered each of the following except:
A   the ecology of plants and animals.
B   genetic theory.
C   the morphology of living organisms.
D   organisms' geographical distribution.
E   the structure of fossils.

Answer B
Genetic theory (option B) was most definitely not considered by
Darwin because he worked on his theory of evolution during the
first half of the nineteenth century, publishing *Origin of Species* in
1859. Genetic theory, however, had its origins in the work of
Gregor Mendel which was only begun around 1856 and not
completed until much later. The term genetics was not coined by
Bateson until 1906. Darwin had therefore formulated his theory
of evolution long before the genetic theory was proposed. All the
other aspects listed were considered by Darwin, in part at least, in
formulating his theory.

Example 2 (Time allowance 1 minute)
Which of the following human characteristics is the best
example of continuous variation?
A   blood groups
B   eye colour
C   the ability to taste the chemical phenylthiourea (PTU)
D   the height of adults
E   ear lobe shape

Answer D
Continuous variation is where the variation between individuals
in a group forms a gradation. Although variations between
individuals at opposite ends of this gradation may be great, there
are a large number of intermediate types each of which is only
very slightly different from the others close to it in the gradation.
Only option D, height of adults, shows such a gradation in
humans. Blood groups (option A) fall into a few clearly
recognisable groups (the most common groupings being A, B, AB
and O). Likewise eye colour, while having some intermediate
varieties, e.g. green, blue/green, nevertheless falls into a few
major colour groups. The ability to taste PTU (option C) classifies
humans into only two groups, tasters and non-tasters, and is

151

clearly not an example of continuous variation. While the shape of ear lobes (option E) is rather various, they can again be classified into a few distinct groups which form far less of a continuum than does the height of adults.

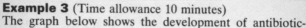

Example 3 (Time allowance 10 minutes)
The graph below shows the development of antibiotic-resistant strains of bacteria at a large hospital over a period of ten years.

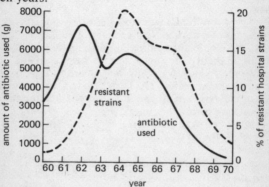

Show how the graph may be used to demonstrate evolution in action.

Answer
According to the Darwinian theory of evolution a variety of organism more fitted to the prevailing conditions and circumstances is more likely to survive and produce offspring than another less suited variety. That is, a selection pressure occurs which favours the survival of one type. In the graph above there can be assumed to be two basic groups of bacteria, only one of which is resistant to the antibiotic being used. Administration of the antibiotic will destroy all susceptible (normal) strains of the bacteria, but not the resistant forms. It is hence the antibiotic itself that provides the selection pressure—selecting in favour of the resistant strains. These strains, presumably mutant varieties of the normal forms, will therefore survive and as they contain the gene(s) for resistance to the antibiotic, these genes will be passed to the next generation. These mutants survive to reproduce their own kind and have the potential for further mutation into other resistant types. The greater the quantity of antibiotic used, the

152

greater the selection pressure and the more rapidly resistant strains develop. From the graph it can be seen that the amount of antibiotic used in the hospital more than doubled between 1960 and 1962. As it takes some time for the mutant resistant strains to build up numbers, and the susceptible strains to be destroyed, the peak in the percentage of resistant strains occurs about two years after the peak (i.e. in 1964) in the amount of antibiotic used (in 1962). The graph of the number of resistant strains thereafter resembles the graph of the amount of antibiotic used, though after a time interval of two years. New varieties of bacteria have evolved which are better suited to surviving the prevailing conditions. In time they may become sufficiently distinct from the original form to be considered a separate species.

Example 4 (Time allowance 25 minutes)
The peppered moth, *Biston betularia*, exists as two varieties: the normal variety, which is a light colour, mottled with black and the melanic form which is totally black.
In an experiment, marked individuals of each variety were released in both an industrial and a rural area. Some of the moths were later recaptured and the number of each variety recorded.
The results were as follows:

Situation	Industrial area		Rural area	
Variety of moth	Normal	Melanic	Normal	Melanic
Number of marked individuals released	120	300	200	250
Number of marked individuals recaptured	30	165	36	15

(a) Compare the results for the two areas and explain the differences between them.
(b) Indicate briefly how two species of the moth could in time develop as a consequence of natural selection.

153

Answer

(a) The number of each type of moth released varies in each area. To make meaningful comparisons therefore, it is first necessary to calculate the percentage of each type recaptured. This is calculated in the following way:

$$\% \text{ recaptured} = \frac{\text{number of marked individuals recaptured}}{\text{number of marked individuals released}} \times 100$$

% recaptured therefore:

Industrial area		Rural area	
Normal	Melanic	Normal	Melanic
$\frac{30}{120} \times 100$ $= 25\%$	$\frac{165}{300} \times 100$ $= 55\%$	$\frac{36}{200} \times 100$ $= 18\%$	$\frac{15}{250} \times 100$ $= 6\%$

When comparing the numbers recaptured in each area it is clear that a greater percentage of both normal and melanic forms are recaptured in industrial areas. The moths must therefore face greater problems of survival in rural areas, possibly due to more predation from the large variety of wild life especially birds in the rural area. In industrial areas the smoke from the burning of fossil fuels in factories and homes contains carbon particles which blacken buildings and trees. At the same time the sulphur dioxide in this smoke destroys many lichens because the algal part of them is readily killed by sulphur dioxide. The survival of the melanic form is markedly greater in the industrial area, where its dark colour makes it less easily seen by potential predators, against the smoke-blackened, lichen-free buildings, walls and tree trunks. The lighter coloured normal variety is particularly conspicuous against such a dark background and is therefore more readily eaten by predators.

Comparing the numbers within each area it is clear that the melanic form survives more than twice as well as the normal form in the industrial area. The normal form, however, is three times more likely than the melanic form to survive in the rural area. Here the melanic form is more conspicuous against the light coloured lichen-covered trees and walls whereas the mottled normal form is ideally camouflaged against such a background. Predation is hence largely confined to the melanic form.

(b) In the industrial area the melanic form presumably arose as a mutation of the normal variety. By chance, the blackened, lichen-free surroundings made this variety less conspicuous than the normal one. It is therefore less readily preyed upon. More of

154

the melanic than the normal form survive to maturity, mate and produce offspring, which are in turn black and better able to survive. This natural selection of the better suited form also occurs in the rural area, only here it is the normal variety that is better camouflaged, less preyed upon and therefore survives to perpetuate its own kind. In time, the less well adapted variety dies out in each area, leaving only the better suited one. Each form is now more or less isolated from the other and apart from the boundary between the two groups there is little or no interbreeding between them. Each form develops its own gene pool, more or less isolated from the other. Further mutations of each form alter the composition of each gene pool and the two become increasingly different. These differences could result in the two forms becoming morphologically, physiologically or behaviourally incapable of reproducing with each other. Two species will then have developed from the original one.

Chapter 15 The environment

Example 1 (Time allowance 20 minutes)
In an experiment to investigate the breakdown of humus by soil organisms, leaves of a known weight were placed in wire cages of varying mesh size. These cages were carefully buried in the soil and over the course of a year the samples were reweighed. The results obtained are shown below:

Month	Weight of leaf samples (g)	
	Sample A (1 mm mesh)	Sample B (5 mm mesh)
January	100	100
February	99	97
March	97	94
April	94	86
May	89	75
June	81	59
July	71	39
August	60	21
September	53	10
October	48	6
November	45	4
December	44	3

(a) Draw a graph of these results.
(b) Give a full explanation of the results obtained.

Answer
(a) In drawing the graph follow the guidelines given earlier on pages 89 and 90. The finished graph should appear similar to the one below:

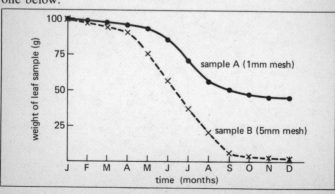

(b) The shape of the graph is similar for both samples. The weight of leaves decreases throughout the year as they are broken down by a variety of organisms. As the largest mesh size is 5mm and no endothermic animal is this small, all such breakdown must be duc to cctothcrmic animals. This accounts for the slow rate of breakdown during the months January to April, when the colder temperatures reduce the rate of metabolism of ectothermic organisms, many of which are in any case dormant over the winter period. During the warmer spring temperatures these organisms become active and so the rate of breakdown increases and the weight of leaves decreases more rapidly. This continues throughout the summer until the rate of breakdown starts to decrease during the cooler autumn temperatures.

Where the mesh is larger (5mm) in sample B, the rate of leaf breakdown is considerably more rapid than where the mesh is smaller (1mm) in sample A. This is because a much greater range of organisms can enter the wider mesh of the cages around sample B. Many of the organisms will consume the leaves. In sample A for instance most decay will be due to micro-organisms such as bacteria and fungi. Sample B in its 5mm mesh cage, will not only be consumed by micro-organisms but also small insects (e.g. beetles), arachnids (pseudoscorpions), crustacea (woodlice) and even small earthworms which may drag out the leaves. The weight of leaves in sample B therefore decreases more rapidly than in sample A.

Example 2 (Time allowance 8 minutes)
The table below shows the amount of DDT measured in parts per million (ppm) found in a variety of organisms associated with a large freshwater lake.

Where the DDT level was measured	DDT ppm
Water	0·0003
Phytoplankton	0·006
Zooplankton	0·04
Herbivorous fish	0·39
Carnivorous fish	1·8
Fish-eating birds	14·3

(a) Calculate the concentration factor from water to herbivorous fish.
(b) What principle is illustrated by the data?
(c) Briefly explain the reasons for the change in DDT levels in the different organisms.

(a) The level in the water   = 0·0003
The level in herbivorous fish = 0·39

Concentration factor = $\dfrac{0·39}{0·0003}$ = 1300 times

(b) The data illustrates concentration of DDT along a food chain.

(c) The DDT level in the water is relatively low. Some diffuses into the phytoplankton where it remains, accumulating especially in any fatty material. Zooplankton feed on phytoplankton. During their lifetime they consume many times more than their own body weight of phytoplankton, i.e. they consume an increasing amount of DDT. This DDT accumulates within them. As one looks down the table each animal consumes a large weight of the animal above it and hence an increasingly larger amount of DDT. The levels of DDT therefore increase rapidly as one moves down the table (i.e. as one moves along the food chain).

Example 3 (Time allowance 20 minutes)
The number of lichen species growing along a 20km transect from the centre of Belfast was recorded at 1km intervals. The results are presented graphically below:

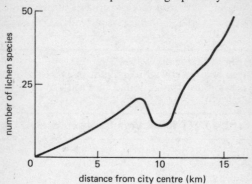

(a) What relationship is illustrated by the graph?
(b) Explain, as far as possible, the relationship between the distance from the city centre and the number of lichen species.
(c) Give one possible reason for the fall in the number of lichen species at a distance of 10km from the city centre.
(d) What is meant by the term indicator species?

158

(a) The number of lichen species increases more or less linearly the greater the distance from the city centre except at a distance of 10km from the city centre where there is a fall in the number of lichen species.

(b) Lichens are especially sensitive to sulphur dioxide in the air, the algal component of the lichen being killed at even relatively low concentrations of sulphur dioxide. Different species show slightly different levels of tolerance. Industrial cities such as Belfast have high levels of sulphur dioxide as a consequence of domestic and industrial smoke produced from burning fossil fuels, especially oil. In older cities such as Belfast the industries and housing are situated in and around the city centre. Sulphur dioxide levels are therefore higher near the centre of the city and diminish slowly as one moves out from it. The pollution from sulphur dioxide is sufficiently large in the centre of the city to prevent any lichen species from growing. As one moves outwards the species more tolerant to sulphur dioxide appear first. As the sulphur dioxide levels fall even more, the less tolerant ones also survive. Finally at some 15km from the city centre around 40 lichen species exist.

(c) The levels of sulphur dioxide determine the number of lichen species and it is therefore most probable that the fall in lichen species 10km from the city centre is due to an increase in sulphur dioxide levels at this point. Such an increase is likely to be due to industrial or domestic smoke from a small town at this distance from the city centre.

(d) This is a species whose distribution is largely determined by some particular factor. The presence or absence of an indicator species therefore gives some measure of the level of that factor. Using a number of indicator species it may be possible to determine the precise level of the factor at one point. One example involves determining the exposure scale, i.e. amount of wave action, on a sea shore using indicator species. In the above case the level of sulphur dioxide pollution could be fairly accurately measured by looking at the number and types of lichen species present. For example if species A is only found where the sulphur dioxide level is less than 2ppm and species B where the level is less than 1ppm, then in an area where species A is present but species B is not, the sulphur dioxide level probably lies between 1 and 2ppm. While other factors do affect distribution of lichens, the use of indicator species nevertheless gives a remarkably accurate indication of the sulphur dioxide level in the atmosphere.

Example 4 (Time allowance 15 minutes)
Consider the following facts about the element strontium.
1   It is a component of human bone tissue.
2   It is readily absorbed by grasses.
3   It is found concentrated in cows' milk.
4   Its radioactive isotope strontium 90 is a product of atomic bomb explosions.
5   Strontium 90 has a half life of 28 years.
Using these facts show why a reduction in atomic explosions could lower the incidence of leukaemia throughout the world.

Answer

Atomic explosions produce strontium 90 which becomes dispersed into the atmosphere. In time, much of this strontium 90 returns to the ground as 'radioactive fallout'. It is readily absorbed by grasses and these account for a considerable amount of the world's vegetation. Grasses form the staple diet of many cattle species throughout the world and cattle in turn are a popular domesticated animal used as a major source of human food as meat and milk. As strontium is concentrated by cows in their milk, much of the strontium 90 absorbed will find its way into humans via the consumption of this milk. As a component of bone it will be accumulated in bone tissue. Young children especially accumulate the strontium because as they grow the minerals are rapidly laid down in their bones and milk forms an important part of their diet. As the marrow cavity of bone consists of rapidly dividing blood-producing cells, this is particularly vulnerable to radioactive damage because:
1   rapidly dividing cells are more vulnerable to mutation.
2   the radioactive strontium 90 is in close proximity to the dividing cells.
3   radioactive materials increase the mutation rate.
One consequence of mutation is to produce an increase in growth of cells in the bone marrow that produce leucocytes (white blood cells). This cancerous growth leads to an increase in the leucocytes in circulation and upsets the balance between red and white blood corpuscles. Such an uncontrolled production of leucocytes is termed leukaemia. With a half life of 28 years, the strontium 90 from atomic explosions can still cause leukaemia a long time after its release and the radioactivity may persist within the body for a long time. A reduction in atomic explosions would therefore, in time, lower the incidence of leukaemia throughout the world.